MRP+
The Adaptation, Enhancement, and Application of MRP II

David A. Turbide CFPIM, CMfgE

Industrial Press Inc.

Library of Congress Cataloging-in-Publication Data

Turbide, David A.
    MRP+ : the adaptation, enhancement, and application of MRP II / by David A. Turbide.—1st ed.
        256 p. 15.6 x 23.5 cm.
    Includes bibliographical references and index.
    ISBN 0-8311-3046-6 :
    1. Manufacturing resource planning.   2. Production planning.
I. Title.
TS176.T87   1993
658.5′03—dc20   92-30393
                                                                                                     CIP

MRP+

Industrial Press Inc.
200 Madison Avenue
New York, New York 10016

First Edition

First Printing

Copyright © 1993 by Industrial Press Inc., New York, New York. Printed in the United States of America. All rights reserved. This book, or parts thereof, may not be reproduced, stored in a retrieval system, or transmitted in any form without the permission of the publishers.

10 9 8 7 6 5 4 3 2

Dedication

To Deb

Wife, Friend, Partner, Cheerleader

Contents

	Preface	ix
1.	Introduction	1
2.	Order Point to MRP to MRP II to JIT to CIM	13
3.	Production Activity Control—Order Based	35
4.	Continuous "Flow" Production	49
5.	Make-to-Order, Ship-from Stock, and In-Between	63
6.	Make-to-Order and Government Contracting	79
7.	Process Manufacturing	98
8.	Activity Based Costing	110
9.	Computer Integrated Manufacturing (CIM)	118
10.	Software: Selection, Adaptation, Modification	131
11.	Measurements	143
12.	Planning for and Implementing MRP II	154
13.	Trends	161
14.	Summary	166
	Glossary	173
	Bibliography	178
	Index	179

Preface

The title of this book introduces a variation or expansion of the acronym MRP II, which stands for Manufacturing Resource Planning. I apologize, in advance, for adding to the seemingly endless string of acronyms that inundate those of us in the manufacturing community who have to try to sort out whether the newest acronym represents a real advance in management theories or techniques or is merely another distraction.

MRP+ is not a replacement for MRP II. In fact, there really isn't an MRP+ at all. Originally, MRP stood for Material Requirements Planning. Next came MRP II which expanded the process to other resources beyond materials, thus, Manufacturing Resource Planning. MRP and MRP-based MRP II systems have been around for more than twenty years now, but today's packaged implementations go far beyond the capabilities of the earlier products. Some contenders don't even use traditional MRP techniques.

The MRP II name has come to mean a rather specific implementation of management techniques and theories; and my intention is to look not only at the mainstream versions of MRP II, but also beyond traditional MRP-based systems.

It's true that MRP-based MRP II, that is, those integrated systems that have Material Requirements Planning at the core, are by far the dominant approach to manufacturing management systems today. It is also true that many people consider this kind of MRP II system to be limited in its ability to address specific industry segments such as custom manufacturing (make-to-order) or the process industries. Additions and variants of traditional MRP II, as well as other nontraditional solutions, have been developed to address the needs of these "other" manufacturing environments.

PREFACE

The purpose of this book is to explore the solutions available and discuss the approaches that are being applied to address the rather unique needs of specific industry segments. Starting with the adaptation of "standard" MRP systems to some of the situations encountered in these industries, and proceeding on to enhanced package approaches and the new alternatives, it is my intention to offer an exploration of the applications and provide some insight into their use.

Few manufacturing companies can be categorized as completely in one industry segment, using only one kind of manufacturing technique. For example, many companies have portions of their production process that are traditional "batch" environments, some continuous flow activities, some blending and mixing, and many other variations. The software solution(s) most appropriate for a given company will likely offer capabilities that address these variations.

The success of the implementation may depend, at least in part, on how well the selected package adapts to these unique needs. I hope to cast some light on the various needs and some of the solutions available to meet those needs.

Software capabilities alone, however, are almost never the primary factors in success or failure of an implementation. Proper systems support can only make the effort easier or more difficult. How the nonsystems issues are handled, namely, leadership, communications, organization, and education, will be the ultimate determinant of success.

Use the information in this book to help make your software selection decisions or better understand and apply the software that you have already chosen; then focus on the important issues. Remember that it is the people that make the difference. Provide them with the proper tools (education) and atmosphere (encouragement and support), and reap the benefits of modern manufacturing management technology.

Outline

The first two chapters discuss the diversity in manufacturing, introducing the basic terminology that will be used throughout the book, and define MRP and MRP II as well as some of the other acronyms and buzzwords in this area of the business: JIT, KANBAN, CIM, World Class, etc. Included here also is an outline of the components of the "core" MRP II system. Systems from different vendors mostly differ in how extensions of this core system are handled. The next several chapters focus on shop floor control functions with information about both order-based and rate-based control systems. Following that, in Chapter 5, considerations for master scheduling (make-to-order versus ship-from-stock) are taken up. Chapter 6 looks at make-to-order manufacturing challenges

PREFACE

and the specific additional needs of government contractors. Next comes an exploration of the process manufacturing environment.

Chapter 8 introduces Activity Based Costing (ABC) concepts, which is an area of manufacturing information management that has received increasing interest, particularly in process industries but in other manufacturing segments as well. Chapter 9 discusses Computer Integrated Manufacturing (CIM) relationships between the business systems and the engineering and plant operations areas. Chapter 10 outlines the steps for selecting a package, and presents the options available for implementing MRP II—adapting a package, modifying packaged programs, or writing custom applications. Chapter 11 looks at measurements and how they can be used with manufacturing management systems. Finally, the last chapters deal with the implementation process, summarize, and look ahead to trends in the manufacturing system marketplace.

1. Introduction

When you talk about manufacturing management (software) systems today, most likely you mean Manufacturing Resource Planning (MRP II). Throughout the 1970s and 1980s, nearly all such application systems were based on Material Requirements Planning (MRP) as the central application, and that is still true.

There have been numerous reports of the demise of MRP II in recent years. After all, how can a computer-based management technique that is more than twenty years old still be valid? The fact is that MRP-based MRP II is alive and well and installed in some 90,000+ sites worldwide.* While there are some alternatives, they are few. They tend to focus on niche markets, and they have not had anywhere near the market acceptance that MRP has enjoyed.

Some of the confusion comes from the marketing hype and glowing trade press articles and books touting "new" theories such as Just-In-Time (JIT), KANBAN, Total Quality Management (TQM), and World Class Manufacturing (WCM). The fact is that none of these is a replacement for MRP II and, in truth, most rely on such information systems as an essential tool for the accomplishment of their objectives. These ideas are defined and explained later in this chapter.

According to the APICS dictionary,* MRP is "A set of techniques which uses bills-of-materials, inventory data, and the master production schedule to calculate requirements for materials. It makes recommendations to release replenishment orders for material. Further, since it is time phased, it makes recommendations to reschedule open orders when due dates and need dates are not in phase. Originally seen as a better way to order inventory, today it is thought of primarily as a scheduling technique, i.e., a method for establishing and maintaining valid due dates (priorities) on orders."

APICS defines MRP II as a direct outgrowth and extension of MRP. Conceptually, however, Manufacturing Resource Planning could be im-

---

*All installed base and market share statistics are 1990 figures, systems from U.S.-based vendors, data supplied by Plant-Wide Research Group, Billerica, MA.

*The Official Dictionary of Production and Inventory Management Terminology and Phrases, Sixth Edition, The American Production and Inventory Control Society, Inc., Falls Church, VA, March 1987.

plemented using techniques other than MRP if they accomplished the same thing, that is, the coordination of all of the company's resources toward the goal of producing the right products at the right time, of good quality, and at an acceptable cost.

The validity of the MRP/MRP II approach has been proven through time at literally tens of thousands of companies around the world. These companies range in size from less than a million dollars in annual sales right up to the top of the Fortune 500 listing, from a few employees to many thousands. MRP/MRP II techniques have been applied successfully in virtually every kind of company, producing nearly any variety of product imaginable from after-dinner mints to massive power generators, from medical catheters to rock crushers.

Yet, despite the apparent universality of the techniques and commonality of needs across the scope of manufacturing, there are distinct differences to be found when comparing plant to plant, company to company, industry to industry. While it is true that every manufacturer needs materials, every manufacturer adds value to these materials in some way to produce a product, and every manufacturer must move (distribute) the product to the customer, it is also true that the specifics of each of these requirements or activities will vary considerably, and the emphasis will change depending on the relative importance or relative scarcity of the resources needed.

A simple machining service operation, the classic "job shop," is very concerned about getting optimum production out of the people and machines that are available but typically doesn't worry too much about planning material acquisition; their raw materials are readily available and are usually not expensive (sometimes they are even provided by the customer).

A company that assembles small electronic devices, which could also be considered a "job shop," might use simple machinery or rely primarily on manual labor. In this case, the availability of components might be of primary concern especially if one or more of the parts is hard to get, is subject to wide price swings, or has a very long lead time. It is likely that materials account for the majority of the cost of goods for this electronics assembler, whereas, in the machining company, labor and overhead could far outweigh material content. Both "job shops" need materials, labor, and facilities, but the specifics of these needs and their relative importance are quite different.

These two job shop examples only hint at the diversity that is to be found in manufacturing. Some companies make only a single (or very few) products, but do so in great volume; others make a wide variety of products. Some manufacturers make only one or a few complex products from a great many parts, while others make a great variety of products from a few basic materials. Some make for stock while others produce only what has been ordered.

Despite this diversity, the coordination of resources that MRP/MRP II strives for still applies. MRP has been proven applicable in virtually all situations, however, mere applicability may not be enough. Some standard "plain vanilla" MRP systems may lack features that allow proper definition of the products or processes or may not support critical needs for a particular industry or situation. In this case, the user must either modify or enhance the packaged software product or develop some kind of "work-around" to adapt the software to the situation. In other cases, although the definition and the techniques apply, there may be alternative approaches that are arguably more appropriate or just plain better.

It is my intention to explore these situations and alternatives in this book. The majority of the discussions herein center around "generic" MRP/MRP II and its derivatives. There is some material about unique or alternative products, but I have generally avoided mentioning vendor names and specific products since I fear that they may be gone or substantially changed by the time this text makes it through the publication process and into your hands.

The Job Shop

The term "job shop," introduced above, is one of those names that has almost as many meanings as there are people who use the term. In the context of MRP systems, I have heard: "MRP is designed for job shops and we're not a job shop. We're..." as many times as I've heard "But we're a job shop and MRP isn't made for a job shop. It's for..." It can all be quite confusing.

Some companies that market "job shop" (non-MRP) management systems have built an entire marketing campaign around the definition of job shop and the requirements of companies that fall into this category. As a marketing tool, this approach can be quite effective, especially since the definition of the term can be manipulated to include or exclude virtually every manufacturer in existence, depending on the aim of the marketing program.

These companies will often equate a job shop with make-to-order (as opposed to make-for-stock). This definition doesn't help us much when we are looking at management system requirements, however, since it only describes the finished goods inventory policy and the related master scheduling time frame. This definition of a job shop could include the small machine shop down the road that makes different parts every day, an assemble-to-order company that has semi-standard products with variations, or an automotive supplier that produces tens of thousands of the same item, week after week, month after month. The only distinguishing characteristic, by this definition, is that production isn't *completed* until and unless the product is already sold. Completed items go right out to the customer rather than to a finished goods stock area from which they are sold after completion.

The management challenges and system requirements for the three examples just presented could be quite different, yet they are lumped together under the job shop label as if they were all alike.

Under "job shop" in the APICS dictionary, you will find a cross-reference to "Intermittent Production." The distinguishing characteristic for this kind of production is that jobs pass through the production process in lots. That is to say, a lot will stay together throughout its processing and will move from one step to the next as a group. Another term often used to describe this situation is "discontinuous" manufacturing.

The referenced APICS definition begins by saying that the plant floor is organized according to function. While this may have been true of most "discontinuous" production plants in the past, it is not necessarily a distinguishing characteristic of intermittent production any more. Many lot-oriented or intermittent manufacturing environments today include portions of the plant that group dissimilar machines together to efficiently perform a related series of operations. Work flow within this grouping, often called a cell or a focused factory, may be continuous, but there is job-order control overall and you will find the work in lot groupings at the input and output ends (moving to and from the cell).

If you accept the APICS definition that job shop means intermittent production, then traditional MRP must be designed for job shops because it is, by its very nature and structure, oriented toward the tracking of manufacturing activity in relation to orders or jobs, which are the mechanism for identifying lots (groups) of production-in-process as they flow from facility to facility. (There are variants of MRP that include plant-floor management functions to address the needs of continuous production, but that's another chapter.) The argument, therefore, that MRP isn't appropriate for job shops because it is designed for make-to-stock is entirely illusory.

In addition to its lack of an accepted definition, the term job shop is troublesome because it doesn't seem to point to any specific situations or needs that we can address in a discussion of manufacturing management systems and their applicability. Unless, that is, job shop is defined as intermittent production. To avoid the term job shop when discussing the needs of companies that produce in lots, I'll use the term "order-based." The existence of an order (manufacturing order, work order, or job) establishes an identity for a quantity of an item and implies that an identifiable "lot" or job order will move, generally as a unit, through the

INTRODUCTION 5

plant, thereby establishing the nature of the tracking and accounting system. All activities are reported to the order and tracked according to the "order" identity.

The term "custom manufacturers" will be used to designate those companies that have no standard products; every job (and product) is different from all others. This is a special case within the make-to-order grouping and is also, most often, a subset of order-based production with some system requirements that are somewhat different from those of a company that repeats a design. If each order (item) is different, there are considerations in the specification and storage of bills-of-material and routings, but the tracking and accounting functions are the same as all other order-based systems... the order is assumed to move through the plant as a unit. Within this grouping you will also find most contract manufacturing and many government contractors.

Not Order Based

Order-based manufacturing accounts for the vast majority of companies and, by exclusion, defines the other category: those companies that do not produce products in identifiable lots that pass through the production process as a group. Turning once again to the APICS dictionary, we find a definition for "continuous production" in which material flow "is continuous during the production process." I interpret this to mean that the product is free to move from one activity (operation or work station) to the next on its own schedule without having to wait for its brothers and sisters to reassemble themselves into that identifiable group. The rest of the APICS definition specifies that production equipment is organized and sequenced according to the steps involved in the production process (although we already rejected equipment arrangement as a definitive characteristic in the other definition, this statement is generally true for continuous production) and that routings are fixed and setups are seldom changed. These characteristics are also not definitive, but are generally true in many continuous production situations.

This path seems to be leading us toward "flow" manufacturing with production lines and assembly lines. The APICS definition for flow manufacturing focuses on an uninterrupted material flow but allows for a variety of products to be produced. Each product, however, utilizes the same production processes in the same sequence and varies only in material content, e.g., colors and flavors. In addition to bulk manufacturing situations such as for consumer goods (shampoo, breakfast cereal), the flow description can also apply to other high-volume situations such as automobile parts and consumer electronics items.

The last sentence of the flow definition specifies that "Production is set at a given rate, and the products are generally manufactured in bulk." I think we have now stumbled onto the basis of a distinction that is significant from a planning, scheduling, and control standpoint. In this kind of manufacturing, materials flow at a scheduled rate through a relatively fixed process rather than individual rates and sequences which may vary between products. In further discussions, this will be called "rate-based" manufacturing to be distinguished from order-based production. In rate-based production, no order, as such, is identified. Production reporting is tied to the scheduled production by item, by production line, by day or shift.

PROCESS MANUFACTURING

A type of manufacturing that is often found in high-volume production plants is "process" manufacturing, as commonly used in the food, chemical, and pharmaceutical industries. "Process" is usually associated with component materials that are powders or liquids, and processes that involve mixing, heating, and/or chemical reactions. Since process manufacturing can be done in batches (kettles, mixing vats) or on a continuous flow line, either rate-based or order-based management may apply. In fact, it is not unusual for process plants to include batch (order-based) mixing and flow (rate-based) packaging for the same product. Although process products are often made for stock rather than to a custom order, this is not always the case. Process industries do have some needs and orientations that are considerably different from traditional "discrete" manufacturing, and there are variants of standard MRP systems that are aimed at this segment of the market. There are also some nontraditional (non-MRP) approaches to manufacturing management specifically designed to meet the needs of the process environment. Chapter 7 is dedicated to the considerations of this industry segment.

Types of Manufacturing

In summary, the major distinction that we have identified in the work flow or shop floor area is order-based versus rate-based manufacturing. Traditional MRP II is oriented toward order-based environments, but rate-based scheduling and control can be accommodated with variants or enhanced versions of standard systems or through nontraditional approaches.

Regardless of the basis of shop scheduling and control, companies may build to order, ship from stock, or start building products based on forecasts but complete the products to customer order. The major impact of this variable is finished goods inventory policy, but there are also

INTRODUCTION

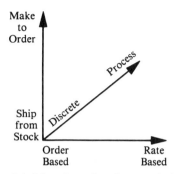

Fig. 1-1. Manufacturing characteristics.

serious considerations in master scheduling, management approach to inventory policy, work-in-process control, and product definition.

The third area of classification discussed above includes the characteristics of the manufacturing processes and materials: process manufacturing which usually involves powders and liquids and often includes mixing, heating, and chemical reactions in the process, as compared to "discrete" manufacturing which is the more common fabrication and/or assembly situation.

The distinctions just outlined are not, in themselves, significant in that they are merely titles and don't necessarily define specific needs or methods. These categories are useful, however, if they help point out variations in the application of management systems and techniques, and they can be used to enhance communications as we discuss these throughout this text.

The diagram in Figure 1-1 illustrates that there is a continuum within these distinctions and that a manufacturing situation will fall somewhere within the three-dimensional grid.

At the lower left end of the graph is the typical end of the industry that the most basic MRP system was first designed to address. Basic functions of material availability planning and accounting, resource availability planning and activity management, and associated support functions are addressed, to a greater or lesser extent, by off-the-shelf packaged software products. The distinction by which further discussion is based will be how far in each direction a particular package extends to offer necessary functions and conveniences to fit the needs of a specific company.

This figure will serve as a guide for the chapters that follow. Each topic will be addressed in terms of these industry characteristics, the methods used to satisfy the needs, and other considerations of interest in light of the three variables illustrated.

Does It Matter?

Despite the fact that this book is dedicated to discussing the features and functions of software packages, the capabilities of the system are seldom, if ever, the prime determinant of success or failure of an implementation effort. Many companies have been spectacularly successful in manufacturing system projects with poor software, limited software, or somewhat inappropriate software. At the same time, others have failed to meet expectations despite the best, most complete, or most appropriate systems for their business.

The obvious question is: why? The answer must be those things that are a part of the effort but are not a part of the system. My experience as a consultant to many such projects has led me to conclude that these other factors can be grouped into four areas: leadership (executive commitment), communications (the team approach), organization (managing change), and education—all "people" issues.

Many companies wrongly assume that they can buy MRP II. Although there are a number of vendors that will be happy to sell MRP II software, success in its implementation takes hard work and dedication within the company. Since considerable effort and resources are required, senior management must be involved in allocating those resources and setting direction for the project. The chief operating officer (president, general manager, plant manager, vice president of operations, etc.) provides the vision and encouragement.

Since these systems cross functional boundaries, one typical result is that departments that traditionally distrust each other, never communicate, and often compete for resources are forced to cooperate for the good (and sometimes the very survival) of the company. Using a team approach to implementation fosters the cooperation and coordination necessary to bring these people together.

Introducing a new system, and the new way of business that will result, can be scary for the employees in the company. There is a natural reluctance to change, to let go of what has worked in the past, no matter how poorly, and embrace an unknown. It's a case of choosing the devil you know rather than the devil you don't know. The best way to overcome this fear and reluctance is through education. The only way people will work wholeheartedly for change is if they understand what they are being asked to do, see what's in it for them (that it will improve their lives, make them more effective, help preserve their jobs, help the company survive and grow), and have the necessary tools (knowledge) to participate.

Change affects the organization as well as the individuals. A change in the management approach or system almost always includes changes to the organizational structure, the lines of communications, and the motivational systems that are applied. In order to be successful, senior management must recognize the need for change in the organization and manage that change as the implementation project moves ahead.

Education is the best investment you can make. Surveys have shown, time after time, that those who invest more in their people garner proportionally larger returns on their system implementation investment. Looking back, nearly every company, successful or not, will admit that one thing they would change, were they to do it over again, would be to provide more education and training.

Finally, a few words about consultants. Consultants can't sell you success, either. A consultant can warn you about, and thus help you avoid, errors and problems that he or she has experienced elsewhere. A consultant can help you get organized, schedule and manage your project, and can assist in the education of the users. A good consultant can help you get the most out of your system. A consultant cannot and should not take responsibility for any part of the implementation effort. It is your system, a tool that you will depend on long after the consultant is gone. Your company and its employees must take an ownership interest in the new system.

What Software Should Do

A computer system—hardware and software—is a tool that people can use in the day-to-day pursuit of their duties. Computers are not a substitute for management judgment and should not be expected to make decisions. They should not take the credit for success, neither should they take the blame for failure.

Some software systems attempt to cross over this line into the realm of decision making, or at least it can seem so if you believe the marketing hype.

One of the most sought-after areas of computer research is that of artificial intelligence (AI). Today, that means rule-based programs called "expert systems" that perform "logical" operations by trading off a number of factors and arriving at a conclusion in what is hoped to be a simulation of human reasoning. This technology can be effectively applied in helping humans stay within the bounds of predefined limits.

Expert systems attempt to codify the reasoning process of knowledgeable people in a particular area and offer the less-than-expert user the advantage of having this advice at hand. The machines do not think, however, and the user should understand that he or she is still the one that makes the decision.

Available packaged software products, including some in the manufacturing management area, are beginning to feature expert systems in such areas as configuration management and finite loading (scheduling) logic which is discussed in some detail in a later chapter.

Expert systems notwithstanding, the computer should not be expected to make decisions. Systems, at best, can turn data into information. If that information is presented effectively, the user can make informed decisions which take into consideration the impact of the proposed action on every other area of the company. The magic of integration allows a user in one area to benefit from the participation of the users in every other area of the system and vice-versa.

The price of admission into this world of information is the full and enthusiastic participation of all functional areas within the company. As in the old cliche, the chain is only as strong as its weakest link. The benefits from the system are limited by the completeness of participation by the users in all areas of the company.

Who Has It, Who Doesn't

I mentioned that there are in excess of 90,000 installed systems out there, and it may seem that everyone who needs one already has it. This is by no means the case. Looking at the U.S. figures alone, there are an estimated 550,000 manufacturing companies and, with only 65,000 systems installed, that is a market penetration of only 11%—nearly half-a-million companies still don't have an integrated business and planning system.

The penetration rates vary greatly by industry. Approximately 44% of the electronics industry has (MRP II) systems, as does 60% of automotive, 70% of aerospace and defense (prime contractors), and over 80% of the pharmaceutical industry. On the other hand, there are nearly 200,000 "other" process industries (not including food, chemical, beverage, and pharmaceutical) with fewer than 1000 systems installed. (See Figure 1-2.)

To say that MRP II is dead or that the market is saturated demonstrates a lack of understanding of the size and potential of this opportunity. For manufacturing companies to compete effectively in a world of increasing competition, they must be able to manage information. Packaged MRP II systems offer readily available solutions that have been proven, in thousands of companies, to be effective management tools.

By the way, it is also interesting to note that the majority of MRP II systems are installed on mid-range computers (minicomputers). Fully 70% of the installed systems in 1990 were on mid-range hardware, and the percentage is increasing as mainframe users are taking advantage of the lower price/increased performance of the smaller systems. Microcomputer-based solutions actually lost market share in 1990, with their growth rate lagging behind the growth of the total installed base.

INTRODUCTION

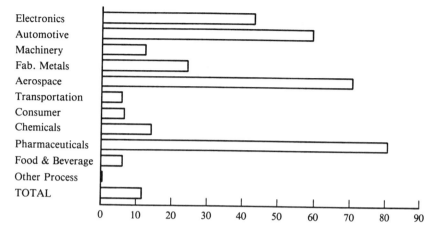

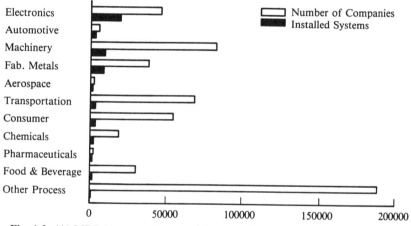

Fig. 1-2. (A) MRP II penetration by industry. (B) Number of installed systems and number of companies.

What Else is There?

Although MRP-based MRP II is by far the dominant solution for manufacturing, it is not the only choice. Other solutions have been developed, in addition to the MRP derivatives, to address particular industries or support alternate philosophies. Where appropriate, I have included discussion of some of the more interesting alternative approaches in the sections where their uniqueness applies.

You will find a discussion of finite scheduling logic in the chapters on shop activity control as a variation on the MRP theme, and you will also find a reference to optimization techniques as an alternative approach. In the chapter on process manufacturing, the MRP enhancements that support this industry's needs are presented, as is a description of a unique (patented) approach used in a product developed specifically for process manufacturing.

Other techniques may exist with which I am not familiar or have chosen not to address, and I'm sure that individual companies have developed their own "brand" of manufacturing information management system that would be of interest; but I have limited the scope of this book to those techniques that are currently widely recognized and used. The future may bring new ideas into this marketplace. Software products and management theories are particularly prone to "better mousetraps" appearing with alarming regularity.

MRP in its various incarnations has been around for more than twenty years now and has yet to be displaced by an indisputably superior approach. I believe that it, and its derivatives, will continue to be the dominant technique in the foreseeable future. And, considering the low penetration rates of integrated information management systems, there is plenty of opportunity for the current software vendors to leverage the investments they have already made in the development and enhancement of MRP-based MRP II.

Keep in mind, always, that the software is almost never the determining factor in success or failure. Choose your software wisely, but once it is chosen and installed, continue to focus on the really important factors of the implementation—those that involve the education, direction, and motivation of the people behind the system.

This chapter includes a brief historical perspective of the development of manufacturing management techniques and ideas, starting from manual inventory control through the application of automation to inventory, and the development of more sophisticated, more comprehensive systems and philosophies. When viewed historically, the developments follow each other in a logical sequence, from the very simple to the gradually more complex and all-encompassing; from the tracking of status, to the use of status information to develop strategies to compensate for problems, to the application of system information as a competitive weapon and a basis to support continuous improvement efforts.

2. Order Point to MRP to MRP II to JIT to CIM to...

Manual Inventory Tracking (Order Point)

Before the application of computers to the tasks of manufacturing management, everything was tracked manually, if at all. Perhaps the first area that was tracked with a formalized system of manual records was inventory balances. Everyone is familiar with a simple card system (Figure 2-1) on which the beginning balance is posted, then successive receipts and issues are recorded along with the resulting new balance. In such a system, there are two major alternatives for replenishment.

In the simplest method, a new supply is ordered when the balance reaches zero. Obviously, unless the resupply is instantaneous, there will be some period when the item will not be available, usually an unacceptable situation (Figure 2-2).

The second method attempts to compensate for this problem by setting a minimum inventory balance at which the resupply is initiated. In this way, the remaining supply can be used while the resupply activity is taking place. As long as the trigger level—the reorder point—is high enough, there will be enough stock left to last through the entire replenishment cycle (Figure 2-3).

Setting the reorder point or, more simply, order point, is the challenge to be addressed. The obvious solution is to estimate usage per day and multiply by the replenishment lead time in days. If usage is 100 units per day and lead time is 10 days, then the order point should be set at 1000

ORDER POINT

INVENTORY CONTROL CARD

Part No.: 27849-AB-294
Description: FRAMISTAN
Class: PF 27
Value Class: C
Cost: 3.05
Vendor: ACE
Location: 24LB7
Minimum: 100
Maximum: 400

Date	Qty.	Order No.	Balance	Date	Qty.	Order No.	Balance	Date	Qty.	Order No.	Balance
1/2		PHYSICAL	386	ordered 2/21	400						
1/7	39	M28743	347	2/27	60	M70068	16				
1/18	41	M31294	306	2/28	+400	P04716	416				
1/26	130	M35661	176	3/5	90	M81802	326				
2/2	140	M40062	36								
order 250 2/4											
2/7	36	M44629	24 B/O								
2/10	+250	P03874	250								
2/10	24	M44629	226								
2/16	100	M55661	126								
2/19	50	M55794	76								

Fig. 2-1. Inventory control card.

ORDER POINT

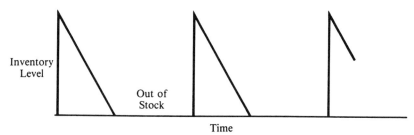

Fig. 2-2. Simple inventory control.

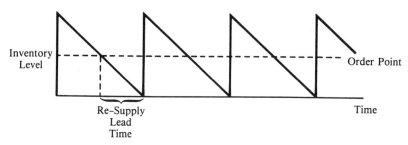

Fig. 2-3. Order point inventory management.

units. When the stock on hand drops to 1000, order more immediately and you should have enough to carry you through until the replacements arrive. The amount to order each time can be calculated using an Economic Order Quantity (EOQ) equation that attempts to balance the cost of processing an order (higher with smaller order quantities because there will be more orders) to the cost of carrying inventory (higher with larger orders because it takes longer to use up the supply). Figure 2-4 contains the equations for order point and EOQ.

There are two factors in the order point equation—usage and lead time—both of which are subject to fluctuations. Usage is almost never constant, day after day, and supplier lead time is not always known, exactly, ahead of time. Even if lead time were known, however, supplies do not always arrive on time, whether from vendors or from your own production facility.

$$\text{Order Point} = \frac{\text{Annual Usage} \times \text{Resupply Lead Time (Days)}}{\text{Number of Days per Year}}$$

$$\text{Economic Order Quantity} = \sqrt{\frac{2 \times \text{Annual Usage} \times \text{Cost per Order}}{\text{Carrying Cost (\%)} \times \text{Unit Cost}}}$$

Fig. 2-4. Order point and economic order quantity equations.

In most cases, the daily usage figure would be derived from an average. Average means "in the middle," and it is safe to assume that half of the time usage would be higher than the average, and half of the time it would be lower. Over the replenishment lead time, we can expect these fluctuations to occur randomly and therefore could expect to run short, i.e., use up all of the stock before the resupply arrives, about half of the time; and that doesn't even consider the times when the resupply is late.

To compensate for these effects, order point users customarily add a buffer to the reorder trigger level. The effect of this buffering is to raise the minimum stock point at which the reorder cycle begins. The buffer amount is commonly called "safety stock." The average inventory level for this situation is safety stock plus one-half of the order quantity as shown in Figure 2-5.

One problem with order point is that the trigger for resupply activity is always based on past events. We reorder when the balance on hand reaches a certain level. That level is reached as a consequence of past activities (receipts and issues). There is no easy way, in order point, to reflect changing usage. If usage is increasing over time, our resupply activity will always lag behind the trend, leaving us more vulnerable to shortages than we might assume. If the trend is down, our supply policy will be unnecessarily high, even if we constantly adjust our order point and safety stock levels, which most companies do not do on a regular basis.

In order to apply some science to this process, we can measure the fluctuations in the demand and use statistical analysis to set order point and safety stock. Some systems will do this (adjust these factors) every month, but remember that we are still working with data from the past. In this statistical approach, the safety stock level is set according to probability theory based on a desired service level. Once the character of the demand is known, the computer can perform the math and come up with the quantity of safety stock needed to achieve a given service level as illustrated in Figure 2-6.

Inherent in the definition of order point is an assumption of risk. We must define a desired service level and buffer to meet it. We can't afford

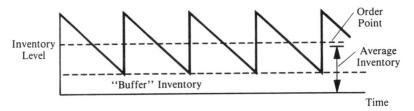

Fig. 2-5. Order point with safety stock.

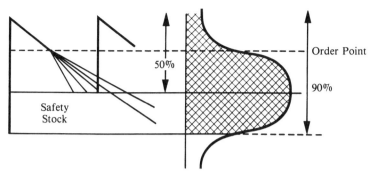

Fig. 2-6. Statistical order point.

100% coverage. As shown in Figure 2-7, the cost of higher service levels increases geometrically. There will be a point above which we cannot afford to buffer. If we determine that we can afford 90% service level, for example, then we are making a decision to live with a 10% probability of running out.

This is acceptable or at least necessary for managing "independent demand" items, those that we sell, since we cannot predict demand with 100% accuracy. We develop the best forecast that we can, measure its accuracy over time, and buffer to the desired or affordable service level.

The service level assumption is a real problem with manufacturing components, however, since more than one is usually required at the same time to meet a need. When we go to the stock room to retrieve enough parts for a manufacturing activity, the probability of all parts being available in sufficient quantity is the product of the probabilities of each individual part's availability. If there are six components, and each is available at the 90% level, the probability of finding all six parts at any given time is 0.9 to the 6th power, or just over 50–50 (actually 0.53). These are not very good odds when you are trying to complete your production schedules on time (Figure 2-8).

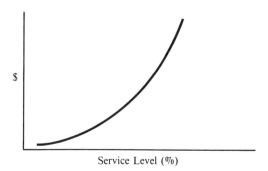

Fig. 2-7. The cost of customer service.

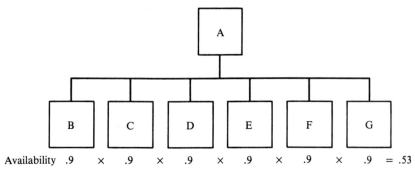

Fig. 2-8. Service level assumption for components.

Material Requirements Planning

Material Requirements Planning (MRP) is the alternative that was developed in an effort to improve this situation. In MRP, rather than managing based on past usage, we look ahead to anticipated need and replenish according to future requirements. We bring in supplies only when needed and only in the quantities necessary to meet the need.

MRP is a calculation technique that uses a master schedule, which is a build plan for the items that you sell, as a starting point. The master schedule is developed from a forecast of demands, backlog of orders with future ship dates, or a combination of these. Like MRP, Master Scheduling identifies expected shortages and plans the activities necessary to prevent the shortage. The master schedule is planned production (start date, due date, quantity) generally only at the sellable-item level. Once the master schedule is developed, MRP is used to plan the acquisition of all of the assemblies, subassemblies, components, and materials necessary to meet the master schedule and all supporting production.

There are four easy steps to MRP. First, the gross material needs are determined by referring to the single-level bill-of-material and multiplying the component quantity-per by the planned production quantity. Next, this quantity (the quantity of each component needed to satisfy the production requirement) is compared to the expected availability of the component items on the date of need. When shortages are detected, the third step is to determine the best quantity to make or buy to fill the need, using a variety of lot sizing rules (or none at all, at the user's option). The fourth and final step is to determine the start date for the acquisition activity by subtracting the component's lead time from the need date.

As you can see in Figure 2-9, MRP is a backward scheduling process. Activities are lined up according to their need dates, working back in time from the planned availability for shipment of the end item, through the bill-of-material, identifying the dates to start acquisition in order to

ORDER POINT

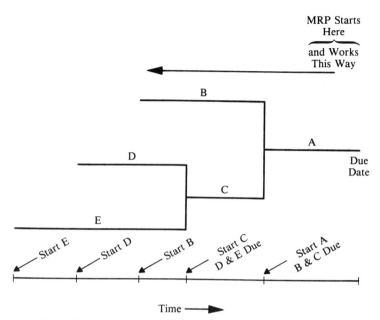

Fig. 2-9. MRP back-schedules from due date to start date.

have the parts available when needed. The result is planned purchases and production activities, designed to meet the need and avoid shortages.

Figure 2-10 presents this process in another format, this time from top to bottom. As you can see, multiple levels of a bill-of-material are addressed sequentially until all levels (all activities and all components and materials) are planned.

One advantage of MRP over order point is that there is no assumption of shortage. If the planning is done correctly, and the plan is executed completely (planned activities accomplished on time), there should be adequate supplies of all needed items at all levels. Another advantage is that MRP is not based on past activities but is tied directly to a statement of future need. This approach is sometimes called a "pull" technique because components are "pulled" in to satisfy a need. Order point is a "push" approach, replenishing based on past usage. The terms push and pull are often misused, however, and should probably be avoided.

MRP is not magic. There are some dependencies that come into play that can severely limit the probability of success. First, the bills-of-materials must be correct. If the wrong item is identified, the wrong item will be made or purchased. If the quantity needed is in error, the wrong quantity will be planned, resulting in either excesses or shortages. If lead times are not accurate, activities will be initiated too early or too late. If due dates are not met or nonconforming items are rejected, shortages will result.

ORDER POINT

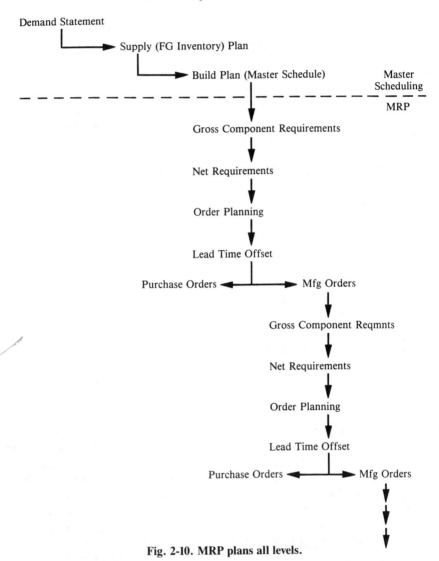

Fig. 2-10. MRP plans all levels.

In theory, MRP can be a true "Just-In-Time" system, in the strict definition of those words. That is, items can be received from their source in exact correspondence to when they are needed. MRP, in truth, seldom works that well because we can't control the data well enough (there are errors in the bills, inventory accuracy is not perfect, etc.) and things do not always go as planned (there is scrap, poor quality causes rejections, orders are late), so we buffer with extra inventory. The more uncertainty there is, the bigger the buffers must be to avoid unpleasant surprises (shortages). Another factor is the lot sizing step. Any change in lot size, above the quantity required, will result in inventory.

MRP is a calculation system that allows us to operate in a less-than-perfect environment. If we know (or can guess about) what our weaknesses are, we can include factors to accommodate them, such as the buffers just mentioned. Even less-than-perfect MRP execution should result in reduced shortages along with reduced inventory levels as compared to order point. Success comes from gaining control of the unknown or uncontrolled aspects of the business, and reducing the buffers that we have built into the planning process to compensate for them.

The primary objective of MRP is to avoid shortages. Often, there is a misconception that MRP is designed to reduce inventory. While inventory reduction is a benefit that often results from successful MRP implementation, the amount of inventory required to avoid shortages is a direct result of the accuracy of the data, decisions made in setting the operating parameters of the system (buffers and lot sizing decisions), and the success in carrying out the plan (meeting due dates and reduction in scrap, rejections, etc.).

The foregoing is to contrast with the idea behind Just-In-Time (JIT), which is a focus on the identification and reduction of waste. There is no specific calculation or software product that defines JIT as there is with MRP. Rather, JIT encourages people to continuously look for opportunities for improvement. Closed-loop MRP is often used as a tool to help identify problems (opportunities for improvement) and compensate for less-than-perfect conditions until they can be fixed.

There are computer programs and (MRP II) application features available which are called JIT systems. What this usually means is that there is an ability within the system to identify those components and/or materials that can be delivered in the exact quantity and at the time of need, and there are special inventory control functions to accommodate these items. This kind of JIT feature is discussed in the chapter that covers continuous production.

The Four Steps of MRP

In the first step of MRP, the system starts with the planned production quantity for the parent item (the master scheduled item in the first MRP level) and uses the bill-of-material to identify the components needed. It is assumed that all components are needed at the start of the production process, unless otherwise indicated. Most systems provide for an override to the production lead time for components not needed at the start of production. The component quantity required is the order quantity times the bill-of-material quantity-per, as illustrated in Figure 2-11.

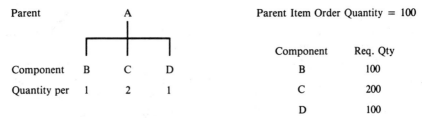

Fig. 2-11. MRP step 1, gross requirements.

This first MRP step can be called "post to components," "bill-of-material explosion," or "gross requirements" generation. Its effectiveness is based on the accuracy of the bill-of-material definition. Systems cannot allow an item to be a component of itself, as this would cause a "loop" in the logic and the program would lock-up or abort. A well-designed system will check for such structuring problems as the definitions are entered rather than wait for the requirements generation to fail.

The second step, called netting (Figure 2-12(A)), is to check availability of the components to identify any NET requirements or shortages. The current on-hand quantity, less expected usage, plus expected receipts between today and the date of need is compared to the required quantity. If a shortage is identified, the planning process continues. If sufficient stock will be available, the process is complete for that component.

This is a time-sensitive process. The system must identify the expected usages and receipts between the present date and the date of need. These include existing acquisition activities (on-order quantities), allocations for existing needs such as production orders already released and customer order backlog, plus activities previously planned but not yet initiated (released). See Figure 2-12(B).

	Component B	Component C	Component D
Required Qty	100	200	100
Available Qty	200	120	0
Net Requirement	0	80	100

Fig. 2-12.(A) Netting—the second step of MRP. (B) Netting is time-phased.

Date	Component C		Available
Today	On Hand Quantity:	250	250
Today	Allocation for Production	75	175
//**	Requirement 1	50	125
//**	Requirement 2	100	25
//**	Expected Receipt	200	225
//**	Requirement 3	105	120
//**	This Requirement	200	−80

To identify all of the planned activities, the system must gather all requirements for an item from all sources before checking against availability. When bills-of-material are entered to the system, the computer will determine a "low level code" which indicates the lowest level on any bill at which this item resides. Requirements are generated down to that level before netting can take place.

Effective netting relies on accurate inventory balances. The dates and quantities of expected receipts and usages must also be accurately represented in the system for the plan to be useful. The fact is that this is often the most difficult part of MRP implementation. Control of inventory movement and the accurate, timely reporting of it is the biggest challenge in MRP because it usually involves many people (stock clerks, production workers, material planners, receiving, purchasing, production control, and many others); it is not exciting or glamorous (report each receipt or issue, when it happens, accurately), and it happens many times each day. Discipline and an "ownership" interest in accuracy are the keys to success in this area.

Also during this second step of MRP, the system will attempt to satisfy any identified shortages by recommending changes to existing acquisition activities. Typically, the recommendation will be to expedite (move a due date earlier) if there is an expected shortage but an order exists with a later due date. An acquisition that is expected when there is no shortage will be flagged for deferral to a (later) date when it is really needed. On-order quantities not needed at all will be flagged for cancellation.

After NET requirements are identified, step three applies order sizing logic to determine the most effective quantity to make or buy. There are several lot-sizing techniques that can be applied, and most systems offer a choice, assignable by item. Among the most common are: discrete (also called lot-for-lot, which orders exactly the quantity needed, actually not lot-sizing at all); fixed quantity (always order X at a time); days of supply (order whatever quantity is needed for the next one week, one month, etc., at a time); or an economic order quantity calculation called part-period-balancing (like EOQ), which compares the higher carrying costs resulting from larger orders against increased ordering costs for smaller, more frequent orders. Most systems also provide for minimum, maximum, and multiple overrides that are applied after any other lot size calculations. See Figure 2-13.

After step three, we have the planned acquisition quantity and the due date. The only remaining piece of information to be determined is the date to start the acquisition activity, that is, to release the purchase or production order or schedule. The start date is obtained, in the fourth and final step of the MRP process, by simply subtracting the item's lead time from the due date. The lead time can include both fixed and variable (lot

Date	Net Requirement	Discrete Order/Balance	Fixed (100) Order/Balance	Days Supply(5) Order/Balance
1/1	50	50/0	100/50	185/135
1/3	70	70/0	100/80	/65
1/4	65	65/0	/15	/0
1/7	80	80/0	100/35	210/130
1/9	40	40/0	100/95	/90
1/10	90	90/0	/5	/0
1/11	35	35/0	100/70	165/130
1/14	60	60/0	/10	/70
1/15	70	70/0	100/40	/0

Fig. 2-13. Examples of various lot-sizing techniques.

size dependent) elements, and must be specified as the "usual" time required to acquire the item. The lead time calculation should recognize only valid work days (skip nonwork days such as weekends and holidays).

One of the chief arguments against MRP is this assumption of a "standard" lead time. It is an accepted fact that lead time is not constant. For purchased items, different vendors may have different lead times, lead times may vary by season, and they often vary unpredictably even from the same vendor. Production lead times are also not likely to be the same every time. The actual lead time will depend on what other work is in the shop, the relative priorities, manning levels, machine down time, etc.

The basic problem is that you don't know what the situation will be when the order is launched. Plans are just that—plans. The advantage of planning systems is that they provide you with the ability to generate the plans and see what the problems are likely to be, and hopefully early enough to do something about it. Using assumed lead times is a means of generating a general view of the expected situation. The shop floor control and capacity requirements planning systems are designed to help you further identify the problems and manage them. This general topic is addressed in more detail in the next chapter dealing with shop floor control and work flow management issues.

Manufacturing Resource Planning

Manufacturing Resource Planning (MRP II) is a natural outgrowth of MRP. Since MRP plans the activities that are to be carried out by such functions as production control, purchasing, and inventory control, it only makes sense to link these applications together (planning, production control, inventory, purchasing) around a shared database so that status information can be easily passed to the planning functions, and so that recommendations can be electronically linked to the release and execution functions. Add customer service and accounting applications and you have MRP II.*

Manufacturing Resource Planning systems are readily available from any of a number of software vendors, and from many of the larger computer hardware vendors as well. These products are usually, but not always, modular, which allows the user to buy only the applications needed, and to have the flexibility to add additional applications at a later time. A few vendors sell their systems on an "all or nothing" basis. Material Requirements Planning is simply one application area within such a comprehensive integrated software product.

"Basic" MRP II

There are a number of features, functions, and basic organizational factors that are common to all packaged MRP II systems. What distinguishes one package from another is primarily the "extras" that are built into the system in addition to the core functions. There are some minor differences in how the basics are addressed, but for the most part these differences are inconsequential. The basic functions revolve around material planning, of course, and include the following.

Product data, which are the definitions that are used in the planning and execution functions.

The *planning* functions, which include master scheduling, forecasting, MRP, and capacity (resource) planning.

Operations, which comprise inventory control, shop activity control, and purchasing.

Customer service functions of order entry and sales analysis and related activities.

In addition, MRP II should also include financial and accounting functions which are not addressed as a separate area in this discussion but are included as considerations in the other areas.

The following sections identify these basic areas and the functions that they are expected to perform. Each will be addressed in more detail in later chapters when differences and extensions are discussed.

Product Data

In order for the system to help plan and carry out manufacturing activities, it must know fundamental information about the products,

---

*Some authorities do not include the financial applications (general ledger, accounts payable, accounts receivable, payroll) in the definition of MRP II.

processes, and facilities. This information is entered into files which are kept current by the user through file maintenance activity. A system should monitor actual experience and provide feedback to the user when reality is different from the maintained definitions (either on a case-by-case basis or through averages, or both), but the "actual" data should not automatically replace the maintained "standards." I have placed the word "standard" in quotes because it tends to be a rallying point for those who would claim that MRP (MRP II) is not appropriate for custom manufacturing situations. These companies, they claim, have no standards since every product is unique. I only use the word "standard" here in the sense that these definitions are entered and controlled by the users and are not the result of actual activity reporting. Perhaps "not-actuals" would be a better term?

In any case, the product data includes both the material relationships, i.e., bills-of-materials, and the process definition: the value-added activities that turn parts into products. The bill-of-material defines the material relationships: raw materials to parts, parts to subassemblies, subassemblies to assemblies, or components to a formulation.

The process definition is often called the routing and is used in conjunction with a definition of the facilities (work centers, work stations, production lines). These definitions are used to develop plans and schedules, establish tracking records in the system when activities are initiated, and serve as a basis of comparison for actual (reported) activity data.

Operations

In this category, the most basic function is inventory control and accounting. The system must accept transactions, keep track of on-hand balance, provide counting support (both total physical and cycle counting), and be able to provide a dollar value for the stock on hand.

In addition, as a basis for the planning applications, inventory accounting must support the availability calculation, tracking expected usage (allocations) and expected receipts.

Also in this category are the execution functions of shop activity control, also called production control, and purchasing. These subsystems accept order recommendations from the planning systems or directly launch (release) production and purchase orders on user entry. They establish tracking records, accept reports of activity, monitor and report schedule information (expected completion, whether the activity is early or late, relative priority), and provide actual cost accounting functions and variance reporting.

Planning

Material Requirements Planning (MRP) has already been defined earlier in this chapter. It plans the release of manufacturing and purchasing activities necessary to supply the materials needed to support the master schedule. In addition, recommendations are made concerning ongoing activities (expedite, defer, cancel) to better align them with the current estimate of need, to assure coverage and minimize inventory.

MRP inherently uses an infinite loading philosophy, which ignores capacity constraints while planning the availability of material (this aspect of MRP is a major topic of a later discussion). Obviously, capacity constraints cannot be ignored, so a followup application—Capacity Requirements Planning (CRP)—helps to analyze the results of the MRP process, as well as the effect of current activity (released orders), and presents the load versus capacity information to the users for problem identification and resolution.

Forecasting could logically fit into either this grouping or under customer service. This process is sometimes a part of an integrated system and sometimes performed separately and fed into the MRP II system. I believe that it is important enough to be a required application of MRP II; and I would consider a packaged system without an integrated forecasting function to be incomplete.

Master Scheduling

While it is an essential part of the planning process, master scheduling is treated here as a separate topic because of its importance, as well as its relevance to at least one of the dimensions of the differences matrix (make-to-order versus ship-from-stock). This area includes the master scheduling function itself, as well as production planning, resource requirements planning, rough-cut capacity planning, and available-to-promise.

Generically, the function of master scheduling is to develop the build-plan (planned production including start date, completion date, and production quantity) for sellable items—finished products and any lower-level items that are sold as parts. Many companies accomplish this task manually or on personal computer (PC) spreadsheets, separate from the MRP II system, but the discussion here will focus on an integrated capability.

Developing the master schedule itself is sometimes a second step in this overall process. Instead of developing the schedule individually for each item, products may be grouped together into "families," groups of items that use similar resources in the production process. The planning of families of items is called production planning.

There are two resource/capacity checking functions contained in this segment, recognizing the critical need to verify the plan before continuing with detailed material planning through MRP. At the family (production planning) level, resource requirements planning does the job. Rough-cut capacity planning provides a check after master schedules for individual items have been developed. Both of these functions focus on what the user has identified as "critical" resources—those that may determine the overall throughput of the plant.

Available-to-promise is an offshoot of master scheduling. This function presents a display or report to the user that shows the planned production (the master schedule) as it compares to the backlog of customer demand. With this presentation, marketing or sales personnel have access to the future available quantities of items that can be sold without interfering with current promises or the master schedule. Hopefully, they will use this information to make realistic commitments to customers.

Customer Service

To complete the picture, there must be a facility to capture customer demand to support the following needs:

* to track the orders to be able to fill them effectively
* to handle credit, shipping, and invoicing functions
* to pass demand information to the planning applications
* to capture demand history for sales analysis and forecasting.

This area also works closely with inventory control and might include distribution management functions. Although this is certainly an important area of the business and the system's usefulness, it is not important to the discussion of the manufacturing management functions that are the subject of this book. No more time will be spent on customer order management except as it may be important to another topic at hand.

Financials and Costing

Again, there is no detailed discussion here about financial applications (general ledger, accounts payable, accounts receivable, and payroll) since they do not directly impact the management issues this book addresses.

Costing considerations, however, are integral to many of the functions described here. Integral is the key word. Costing functions should occur "automatically" as a result of activity reporting and the existence of definitions in the computer files.

There is a heightened interest in costing systems in light of the changes that have taken place in manufacturing since the original "rules" of costing were developed many years ago. When much of the manufacturing value-added was a result of the direct involvement of human hands, it made sense to focus on the labor content as the primary concern and as the basis for assigning overhead. Today, a labor content (cost-of-goods) of something less than 10% is not uncommon. New ways of looking at costs are emerging which address this change. Activity-Based Costing (ABC), the emerging methodology, rates a separate chapter in this book.

Closed-Loop System

The integrated nature of MRP II requires that all subsystems are designed to share information with all others. This is the source of the magic multiplying effect of MRP II. Also called synergy, this multiplying phenomenon occurs when a participant actively maintains his or her portion of the system. Due to the fact that other users are doing likewise, each benefits from the participation of the others.

In effect, one plus one equals three. Each receives more information back from the system than he or she contributes. There is benefit, therefore, for all users.

I would like to reemphasize the scale of importance in system implementation. While this text is concerned with the functions and features of software, it is the quality of the participation by the users that marks the success or lack of success in putting these functions to work for the benefit of the company and all of its employees and stockholders. As each system or function is implemented, it is important that each user has a clear view of his or her role in the process, and how important it is that each user develops a proprietary interest, a sense of ownership, in his or her own portion of the system.

Software and hardware are almost never the prime determinants of success or failure. Inappropriate software, however, can certainly make success harder to achieve or can limit that success.

Just-In-Time and KANBAN

I mentioned Just-In-Time above and used the term literally, that is, to mean delivery exactly timed to need. The JIT that is widely discussed in the trade press does not necessarily mean that, although that might be one result. The JIT that has been widely praised, especially in its successful use by Japanese companies, is not limited to the timing of inventory receipts, but is an all-encompassing focus on continuous improvement and the elimination of waste. There is a popular misconception that JIT can be purchased as software. It cannot. JIT is an operating philosophy.

JIT could more properly be called a war on waste. The interesting thing, however, is how waste is defined as anything that adds cost to the product without adding value. That's a very important definition, as it sets the tone for JIT. By identifying anything that adds cost, we are not limited to inventory considerations alone, but can look to all areas of the company for opportunities for improvement.

Inventory is, of course, an area that offers a lot of opportunity. By the above definition, any inventory can be considered to be wasteful since the existence of stock on hand does not add anything to the product itself. But it certainly has a cost. By this definition, then, all inventory is waste and should be eliminated... thus, just-in-time deliveries. We've already seen that MRP and MRP II strive to reduce or eliminate unnecessary inventory, so MRP and MRP II are compatible and supportive of Just-In-Time. The two are not competitive products at all (remember, JIT isn't even a product). In fact, most JIT programs rely on effective data management and planning systems to help identify the areas of opportunity and to help manage the improvement efforts.

Another term that is widely misunderstood is KANBAN. The term refers to a specific management technique, originally developed by Toyota, in which a physical signal is used to trigger either stock issues or production of replacement goods based on the usage of items in stock. It works in the following way.

The main idea behind KANBAN is actually not a new idea at all but one that goes back to a very simple, early inventory management technique called "two-bin order point." Picture a noncomputerized warehouse in which there are two storage areas (containers or bins) for a production component. One of these containers is located at the point of use, normally at the production line, and the other is in the storage area. Production requirements are satisfied from the "line" container.

When the line container (bin) is empty, the reserve supply is brought in, and refilling of the now empty (formerly active) bin can be started. The existence of an empty bin in the production area is the signal that authorizes replenishment (swapping bins with the stock room). So, in the two-bin system, a physical replenishment signal on the shop floor is used. Contrast this with "conventional" order point in which a minimum stock level is identified either by computer, by an entry in a card file, or by a stroll through the stock room looking for nearly empty shelves.

KANBAN uses the same type of physical signal as two-bin order point. Sometimes it's even a bin, believe it or not. But KANBAN can be more flexible than the two-bin system. In a simple illustration, let's assume that you have five bins, two on the "floor" and the rest in the stock room. A material handler can patrol the shop area, and when he or she sees an empty container, can bring in a full one, avoiding a shortage. You could even have a two-tier system in which the number of empty bins in the stock room is used to trigger reorders from the supplier.

Taking this idea one step further, think about a high-volume situation in which component parts are produced in a different area of the same factory in which they are used. As the parts are produced, they are moved to a staging area near the usage point, each with an identifying tag attached. When a component is "picked" from the staging area for use in the assembly, its tag is passed back to the component production area where it becomes the authorization to build another one. The tag, in essence, becomes the "work order" for replenishment of the component supply.

This simple idea—a physical signal to replenish—now becomes a production control system for components. The number of components to be produced, and therefore the production rate, is determined not only by the usage rate but also by controlling the number of cards or tags (KANBANs) in circulation. To increase the supply and/or the production rate, add cards to the system. To cut back, remove cards. Here is where KANBAN becomes only as effective as the underlying management system. Pure KANBAN, as outlined, can only work well if the production rate is constant or the number of cards is closely managed. KANBAN works best for common, high-usage parts.

Since the production rate (supply) of KANBAN-managed components is determined by usage, KANBAN cannot recognize expected future changes in demand and will result in a component inventory (completed or in process) equal to the number of cards in circulation. KANBAN, therefore, is a "push" system, like order point, and would seem to be incompatible with MRP which is a "pull" system. In fact, however, many companies have combined the advantages of MRP with the expediency of KANBAN for an effective hybrid system.

In such a hybrid, KANBAN is used for common, high-usage components, usually to control production rates for "feeder" lines. The number of cards in circulation is tied to the production schedule for the assemblies. As an alternative to cards, companies also use tags, leather thongs, tokens, or the bins, carts, or trays that hold the parts as the signaling device.

Computer Integrated Manufacturing (CIM)

One of the "buzzwords" that has been very popular in recent years is CIM. Many hardware and software companies have been touting the advantages of CIM for quite some time, while the technology has been quite slow in catching up with the marketing presentations. CIM is an extension of what happened with MRP when it was expanded to MRP II. That is, it became more comprehensive, encompassing a broader range of functions into an integrated whole.

In MRP II, the planning systems are tied into the execution functions and accounting and order management areas. CIM brings in the engineering and shop floor through interconnections that typically were not found before.

One of the big challenges with CIM is that the systems in use in these three disparate areas of the company are typically incompatible. They evolved in different ways, use different operating systems, and adhere to different standards. Equally as difficult is establishing cooperation and communications between the people in these fundamentally different functional areas.

CIM has advanced greatly in the last few years with the development of international standards, accepted proprietary formats, and translation and conversion facilities to allow a relatively painless exchange of data. As CIM evolves in the future, more and more applications will be developed or adapted to emerging CIM standards until true integration allows the interaction of program to program across multivendor applications.

World Class Manufacturing

This is one of those buzzwords that sounds so good but really doesn't tell us much of anything. In today's world, virtually every company is in a worldwide market. Even those that don't export must still be concerned about competition coming from outside the boundaries of their home country. What does it mean to be "World Class?" It probably means that this company can at least hold its own, that is, compete in the global marketplace.

What most writers (and it is really the writers that create these terms, perhaps to sell books, perhaps to promote their consulting services under a "new" flag) mean by "World Class" is actually world competitive.

We can take it a step further, as many authorities do, and talk about being better than your (worldwide) competition in at least one thing, and equal in all the others. You might have a quicker response time (shorter lead time), for example. All else being equal, this could be the competitive edge that allows you to gain market share. Your advantage could be price, quality, variety, almost anything that sets you apart from the rest.

How does one embark on a "World Class" campaign? I don't know. It really depends on what it is you are trying to accomplish—reduce waste, shorten lead times, improve quality, reduce costs, all of the above? A logical starting point is to check out your competition. There is a process widely discussed in the Quality community called "benchmarking." It consists of identifying the best companies there are in your business, analyzing what it is that makes them the best, and emulating them. The remaining step is to become better than the others at just one thing so that you will become the best in the business.

In the manufacturing industry, and probably most others as well, we seem to be always looking for the "silver bullet," the magic answer to all our problems. Every year seems to bring another acronym, another buzzword, that is supposed to be the cure-all to "save" (American, Canadian, British, etc.) manufacturing from the (Japanese, Korean, Taiwanese, etc.) dragon.

I have some bad news: there is no silver bullet. We lie in a bed of our own making. The only real solution is heads-up management. We must understand our products and processes, our people and resources, and our markets. We must use that understanding to make good decisions and intelligent choices. We must "stick to our knitting" and manage, manage, manage.

There are some tools available to help manage the information resource that is so vital to the understanding mentioned above. MRP II is one such tool, though not the only one, that can be effectively applied to assist in this quest. JIT is a philosophy that helps us focus on those things that can make a difference. CIM is a concept and a set of tools that can assist in our efforts to bring our team together for a more coordinated attack against our competition. All of these things can help, but we must not allow ourselves to view these tools as anything more than that—tools. In the words of the late Walt Kelley, as spoken through the immortal Pogo, "We have seen the enemy, and he is us."

Recently I have seen several references in the press to MRP III. Although I don't believe that there is yet a specific definition of what this might be, I fear that it is just another acronym to confuse us and lead us away from the basic management issues with which we must contend. We don't need another miracle cure. We need to better use the proven tools we have available today.

Recent market figures indicate that only around 11% of the manufacturing companies in the U.S. have integrated information systems (MRP II). What are the other 89% doing? How are they able to compete today without the visibility that these tools provide?

In addition, although marketing surveys don't show it, few of the companies that have these systems are fully utilizing them. We buy MRP but we install inventory management. We invest in CIM but spend years trying to get MRP and basic CAD/CAM running effectively. I believe the main reason for this failing is that we tend to underestimate the importance of the human side of the implementation, and thus underinvest in education and training. We fail to adapt our organizational culture to take advantage of this new way of doing business. We fail to take ownership of our systems and make them an essential part of our everyday world.

Packaged Software

The contents of this book are based on the assumption that packaged software will be used. Today, there is little justification for custom development of an integrated manufacturing system. Many packaged systems are available, most are very comprehensive and perfectly functional, and the cost (in both time and money) of development is prohibitive in almost any circumstance that can be envisioned.

Any off-the-shelf manufacturing software package will probably contain more function than any individual company will need at any given time. That shouldn't be a concern, as long as the extra function doesn't get in the way. The price of a package is so much less than the cost of development, even development of a much more limited solution, that there should be no real concern about paying for something you don't need. In many cases, the extra functions are things that you might be able to use in the future as you become more sophisticated or as your business needs change.

It is estimated that custom software costs at least ten times as much as the equivalent function in a packaged product. Further, a custom solution will address the needs of the company at the time development begins and is likely to be inadequate by the time development is completed (if it ever is). One thing that stretches out development schedules is constantly changing requirements. Your business doesn't stand still. Yesterday's solution probably won't satisfy today's needs. A package is likely to contain more function than you will need today, providing room to grow.

As I hope you'll see from the discussions in this book, there are packaged solutions that address the major areas of concern for every type of manufacturing. Where there is a shortfall, there is a solution, an alternative, a work-around, or an extension that is available or can be added on to a packaged product.

If you have the desire, if the people in your company can be brought together in a serious effort of continuous improvement, if you are willing to be open minded and positive, then I am confident that you can implement your chosen MRP II system and become or remain competitive in your market.

Let's now explore the similarities and differences between manufacturing companies and MRP II software packages.

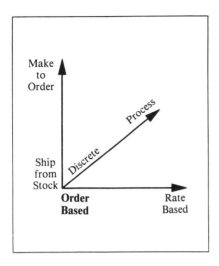

3. Production Activity Control— Order Based

One of the functions central to the management of a manufacturing company is the way work is scheduled, tracked, dispatched, and controlled as it moves into and through the production process. As in all detail areas of this book, we have to recognize the differing needs of the various manufacturing types to make sense of the options and approaches available.

One of the major distinctions called out in the introduction was the manner in which work flows through the shop, i.e., in a continuous flow or a discontinuous, interrupted flow. In discontinuous flow, an "order" for a certain quantity is released to the floor and moves, generally as a unit, through the process. In continuous flow, individual items within a production quantity are allowed to move through the process, each at its own pace. The term rate-based is used here for continuous flow. This chapter will first address the order-based environment. Chapter 4 discusses considerations in a rate-based situation.

Shop Order Release and Scheduling

In an integrated system, the recommendation of what orders to release, their size, and the start and due dates is generated in the planning processes: master scheduling for sellable items, and MRP for dependent items. These recommendations are then reviewed and approved by a planner who will initiate the order through a release function which establishes the existence of the order in the system files, schedules the activities, assigns a priority, and allocates the component materials needed. The applications that manage activity data for inventory and the plant floor will follow through from the release process by allowing the reporting of activities and by maintaining status information, priorities, and actual job costs.

Usually, the release process relies on definitions from the single-level bill-of-materials and the process and facility definitions that are stored in the product data subsystem. All systems allow this information (material and process definition) to be entered directly at the time of release if it is not available in the files or if the user decides not to use the standard definitions; and all systems allow modification of the definitions attached to the order after release, although there are often restrictions on this ability to modify, such as not allowing a change to the standard for something that has already occurred and been reported.

When an order is initiated, it is scheduled (loaded) into the facilities (work centers) that are called out on the process description (routing). This scheduling function can be carried out in either a forward or backward direction, that is, the schedule can be developed by working forward from the start date until the projected end date is determined, or it can be started at the due date, working backwards to determine the necessary start dates for the activities under normal scheduling conditions.

Both forward and backward scheduling are viable techniques, with the choice being mostly a matter of personal preference. Each will provide one or more prioritization schemes that furnish production management with information about the timeliness of the production order (early or late). This priority information is used to manage flow through the shop, typically by managing the sequence in which orders are processed at a given facility.

Traditional MRP II systems carry out this scheduling function using a philosophy known as "infinite loading." This means that the scheduling process is accomplished for each order individually, with no consideration of other load (orders) that may be putting demand on the same facilities in the same time frame.

The facility definition will specify its capacity, usually in terms of hours per shift per day and the number of resources available (persons or machines) in each shift. The calculation for time required for an activity at the facility will be determined by the total number of hours required by the operation step divided by the number of hours available at the facility per day, as can be applied to a single order. For example, if there is one shift and there are three resources available for eight hours, the number of schedulable hours per job per day is eight. It is assumed that three orders can be handled at once, but the system is not limited to scheduling only three orders in the facility in any given day. Since each order is scheduled independently, any number of orders can be loaded into the facility on any given day, assuming that each will receive eight hours of resource.

Infinite loading is not concerned with overload or underload because it is most closely associated with MRP, which focuses primarily on material availability. Once the scheduling is complete, there are functions (capacity requirements planning) that illustrate the effect of the scheduled activities (load) on the available capacity of each facility, but the scheduling process itself is not limited by capacity constraints.

The base assumption in MRP is that the most important consideration is material availability—if you don't have the parts, you can't build the product. Once the parts are scheduled and availability is assured, then the load mismatches can be identified through Capacity Requirements Planning (CRP), and human managers can then make whatever adjustments are necessary to balance the capacity to meet the load, or vice-versa.

If there is an overload, plant management might increase capacity by authorizing overtime, or by adding a person or persons through hiring or borrowing from another facility. They might have some "floaters" that they can schedule into the facilities that need help. They may also consider buying more equipment if the overload is chronic. Another solution might be to reduce the workload by shifting some of the work to another facility, by sending some out to be processed by a vendor, or by changing the schedule.

To have the system make these decisions would require a complex set of rules that describes all of the options, in detail, with priorities and/or tradeoffs between them. While there are systems that do this, the traditional approach in MRP assumes that the capacity situation is better handled through management judgment. The CRP function provides the visibility necessary to support this decision-making process. Don't forget that there is a resource check made at the master scheduling level for a first-cut assurance that the overall schedule is within reasonable expectations. If the master schedule is realistic, mismatches between load and capacity at individual work centers should not be unmanageable.

After MRP develops the schedule for all items contained in the bills-of-materials for all products on the master schedule, Capacity Requirements Planning simulates the scheduling of all of these planned manufacturing activities, using the infinite loading assumption, and presents the resulting load versus capacity information to the user (Figure 3-1). The primary purpose of CRP is to help identify periods *in the future* when resources are either underloaded or overloaded. The user, presumably, will have enough advance warning and the necessary flexibility to rectify the situation before the mismatch can cause missed schedules or inefficiency.

CRP is not intended to manage the flow of work through the shop. It has no capability for sequencing jobs or managing work flow—these

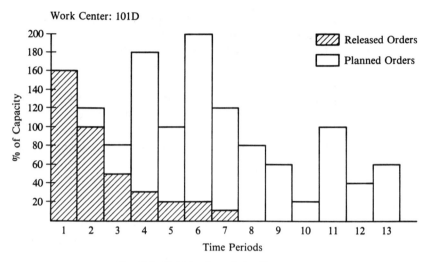

Fig. 3-1. CRP load analysis diagram.

functions are left to the shop activity control or production control application. Some sophisticated CRP modules, however, can offer a simulation of work flow based on the assumption that late orders will move through the shop faster than "normal," and that early orders will be delayed to accommodate this expediting. CRP could then display a "prioritized" or expedited view of the expected load at each work center.

Finite scheduling (or finite loading) is the opposite approach in which capacity constraints are recognized in the basic planning and scheduling process. Finite scheduling advocates contend that it is foolish to ignore capacity constraints "as traditional MRP does." In fact, capacity constraints are not ignored at all, but are considered both before and after the MRP scheduling calculations.

All MRP plans are driven by the master schedule. When the master schedule is developed, identified critical resources should be checked, through rough-cut capacity planning, to determine whether the master schedule is within demonstrated capabilities. If it isn't, the master schedule should be adjusted until the critical resources are not overcommitted.

Later, after the detailed schedules are developed, the overall load, including both released orders as well as planned releases, is checked against the capacity of each facility (work center) through CRP. Any overloads can be addressed, through management judgment, using the options listed earlier. The one basic assumption is that capacity problems can be resolved. If they can't, then the master schedule must be changed to bring it within the capacity available. The user should be able to identify such capacity problems in the rough-cut capacity check during master scheduling. If not detected until the CRP stage, then another iteration (revised master schedule, regenerated MRP, another CRP) is required.

A secondary assumption is that the capacity plan will be studied and any problems (over/underloads) will be solved in advance. If adjustments to load and/or capacity are not implemented, then the predicted difficulties will occur, and schedules will not be met.

CRP can help identify whether capacity problems are chronic or temporary. Chronic problems require long-term solutions—changes in manpower levels, adding or decreasing resource (machine) availability, etc. Short-term nonrepeating problems can often be resolved through schedule changes or temporary measures such as overtime or contracting out.

Queue and Work Sequencing

In a discontinuous production environment, it is typical for 80% of the lead time of a manufacturing order to be nonactive. Of the four elements of lead time—move, queue, setup, and run time—queue is by far the largest component. Queue is the time that an order waits at a facility from the time it arrives, ready for work, until activity actually begins. While an average queue can be measured or computed for a given facility, the actual queue experienced by any given job will depend on the actual workload in place at the time of its arrival and the sequencing of jobs into the facility. If the arriving job has a high priority, for example, it may be moved ahead of its arrival sequence and experience a shorter actual queue time, while all of the jobs that were waiting at the time will have to wait that much longer for their turns.

Managing the sequence of jobs at each facility is the common method for responding to priorities in order-based work flow management. An order that is behind schedule (high priority) should be moved ahead of other, lower priority orders, thus shortening its actual queue and lead time and providing an opportunity to "catch up" and perhaps be completed on time. In a normal situation, as much as 80% of the "normal" lead time for a particular job can be eliminated by priority sequencing. On the other hand, an early order (low priority) can be expected to wait for other, higher priority orders that are moved ahead, thus experiencing a longer actual lead time (longer queue times), delaying it until it meets its due date rather than being finished early.

All of this resequencing results in variation of actual queue times and therefore overall production lead times to accommodate the earliness and/or lateness of orders. As long as the workload input to the facilities is balanced to the capacity (output), the average queue will remain relatively constant. The scheduling system should use a queue time which approximates this average, and the plant floor management will use sequencing to manage the flow of orders to assure on-time completion.

If the facilities are overloaded through the release of more work than there is capacity to accomplish, the priorities of all jobs will increase, and queue and lead times will grow, resulting in late completions. If the inflow of work is less than the outflow, queues and lead time will decrease.

Unfortunately, management tends to address overloading strictly through capacity adjustments rather than by controlling the input rate of work. We tend to drive the manufacturing process to respond to whatever demands are placed on it rather than trying to manage the inflow of work to accommodate the capacity available. Sales and marketing are seldom aware of production constraints and, in the few cases in which they are, there is a tendency to accommodate the customer's requests over any limitations in the plant. Customer commitments are often made within the lead time required to plan effectively. This leads to an unstable master schedule which leads to increased inventory, lengthening lead times, inefficiencies, and expensive expediting. Overcommitment outside of the required lead time will be visible in the rough-cut capacity plan as well as the CRP output, and therefore (at least theoretically) can be addressed before the customers start looking for their (late) shipments. The solution involves training of the sales/marketing personnel to understand lead times and the impact of overcommitment. More discussion of this topic is contained in Chapter 5.

In some manufacturing situations, you don't have the luxury of adequate lead time to plan. This is often the case in the classical job shop or services-oriented company wherein materials are either supplied or are readily available and the chief concern is available capacity of the production facilities. In this environment, material and capacity requirements planning are of marginal interest. What you really need to known is: what resources are available and when? Can I take this job and get it done on time? When can I get this job done?

The desire to plan for and manage production based on capacity has led to the idea of an alternative approach to planning based primarily on capacity constraints. This technique applies the logic known as finite loading or finite scheduling.

Finite Loading

In a hypothetical finite loading system, the planning process would be accomplished primarily according to the limitations of resource capacity, rather than focusing on material availability as in traditional MRP. Once the overall plan is derived, management would then deal with any material shortages that are identified by the system. As you can see, this is exactly the opposite of the traditional MRP approach. Advocates argue that, in many industries, the availability of time on critical production machines or facilities is far more important than any material considerations since materials might be commonly available and/or are a relatively low percentage of cost-of-goods and therefore should not be of concern.

In a custom machine shop, for example, the raw materials could be bar stock, rod, or sheet metal, easily stocked in standard sizes for a relatively modest investment. When a customer order comes in, the most important consideration is likely to be when machine time can be scheduled. Materials don't really matter. If the needed stock is not on hand, it can probably be acquired readily and be waiting in place before machine time becomes available.

Some marketers of finite scheduling software recognize this truth but try to extend it to less idealistic circumstances. Some claim, for example, that capacity limits are the controlling factor for up to 90% of manufacturers.

The truth is that both materials and capacity are important, and no planning function can be complete until both are accommodated. A material-based plan cannot be accomplished on time if it exceeds the resource availability. A resource-based plan will not be completed if there are material shortages.

In a finite scheduling system, the master schedule is used to generate requirements and plans based on facility availability rather than on materials, and the limit of facility availability cannot be violated during the planning process. Thus, as activities are scheduled, the facility load is monitored and, if demand exceeds capacity, the system applies some predefined logic to resolve the capacity problem. In a simple implementation of this technique, the generated solution is to reschedule some activities in favor of others. When all possible rescheduling is completed, any leftover demand (orders that cannot be accommodated) is set aside in a kind of "overflow" area for management to address. Material acquisition is also left for management, under the assumption that materials are readily available or can be acquired as needed. (See Figure 3-2.)

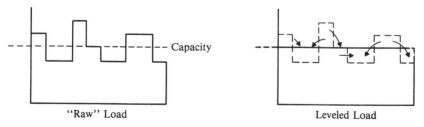

Fig. 3-2. Finite loading.

Some systems use a hybrid approach to finite scheduling. First, a traditional material-based MRP plan is generated, then a capacity requirements plan is developed. In the case of an overload, a series of rules are applied to solve the problem. These rules can include rescheduling orders according to a priority scheme, assuming capacity can be added or shifted within certain limits, or providing management with a list of the things that cannot be done or the recommended changes to the master schedule. Most practical finite loading facilities available today are of the hybrid variety. More detailed discussion of some of the features of these packages is contained later in this chapter.

As stated above, neither technique is inherently right or wrong unless either materials or capacity is the sole consideration, or nearly so. In the vast majority of manufacturing situations, both constraints are important and, as long as both are accommodated, it doesn't make a lot of difference which is planned first and which after.

Conceptually, the traditional MRP approach makes a lot of sense in most circumstances. If there is a material shortage, there are only a few options: take action to bring some in by launching a purchase or manufacturing order or expediting an existing order; or change the schedule for the activity that uses the item that is short (smaller production quantity not to exceed the available parts, start at a later date and expedite to meet the due date, or start later and finish later either delaying final completion or expediting at a higher level on the bill-of-material).

Capacity shortages offer far more options. If labor time is the problem, overtime can be authorized, a second or third shift might be added, weekend work might be possible, a worker might be "borrowed" from another facility, or schedules might be shifted. To solve a machine time shortage, some work could be sent out to a vendor for processing, again another shift or overtime might help, more equipment could be acquired or rented, the work could be shifted to another resource within the plant, or the schedules could be shifted. There are so many options, and the implications inherent in each option or combination of options are so complex, that you may not feel comfortable delegating this decision to the software. It may be more acceptable to let the system lay out the material plan, recommending the few obvious alternatives (which you are free to modify, if you disagree) and you take charge of the more complex capacity balancing.

Traditional MRP does not really ignore the capacity issues or postpone them until the end of the process. As stated before, proper master scheduling techniques will include a check on the availability of critical resources early in the game. By the time the detailed production plan is developed through MRP, there should be no big surprises, although minor adjustments could well be necessary.

The detailed (CRP) planning will provide information as to the resources that are potential problem areas, and a closed-loop approach will use this knowledge to redefine and refine the identification of "critical" resources to watch in future master scheduling activities.

Hybrid Systems

Recently, hybrid systems have begun to enter the market. These systems use classical MRP techniques to develop a material acquisition plan under the infinite loading philosophy. After the initial plan is complete, a finite loading process is applied to schedule the daily activity in the plant. Some of these systems are designed from the start to work this way and are available from the primary vendor; others are add-on packages that can be interfaced to one or more standard MRP offerings.

To implement the finite load process, the user supplies a set of rules that allow the system to prioritize the load to identify which orders stay and which have to go. Another set of parameters is provided (by the user) that the system will use to attempt to solve the problem. These might be something like: add overtime up to two hours per day, then take any remaining overload and shift it up to three days into the future if capacity is available after adding overtime to those days, then displace priority four or less orders from the next time period unless the shifted orders are lower priority, and so forth. Any load that cannot be handled with the rules provided is placed in an exception "bucket" for the user to deal with.

This sounds complex, and it often is. The basic approach begins with the assignment of priorities, usually based on the MRP-derived due dates and traditional prioritizing methods. Other priority considerations might relate to the purpose of the job, for example, orders that are part of the requirements for a customer order might take priority over orders for stock. The finite scheduler then applies the work to the facilities in priority sequence. When the facility's capacity is committed, the rules are brought to bear on the problem. In some systems, the load is developed in the infinite manner, then the situation is displayed with a recommended solution (based on the rules) provided. The user approves or changes the solution using on-screen displays. Others present several possible solutions and ask the user to choose the "best" one.

Many of these infinite loading/finite scheduling hybrids provide slick color graphic displays of schedule and load information which assist the user in the approval/simulation process.

Some finite loading modules provide traceability among parts that are all headed for the same end product. This is especially important if a number of independent activities are all tied to one end item and one of them is delayed. It makes no sense to expedite nine parts through the process if they will only have to wait at the final assembly stage for the completion of the tenth part which was delayed by a machine breakdown or material shortage. By recognizing the interrelationships between jobs (parts) using the actual bill-of-material for the order (it may differ from the standard bill, if there is one), the system can coordinate all of the subtasks to product completion.

The hybrid approach provides the benefits of both methods. Traditional MRP is used to plan the acquisition of materials, set the general parameters of the production schedule, and identify (through CRP) potential future mismatches between load and capacity. The finite scheduling logic supports the short-term objectives of establishing the best sequence and managing the flow of work day to day, recommending solutions based on user-defined priorities.

Not every industry needs the finite scheduling logic. If the production process involves a number of steps and extends days or weeks into the future, traditional CRP and shop floor infinite scheduling are usually fine. If customer commitments are made with only a very few days turnaround, or if the full loading of resources each day with a variety of short-term requirements is the key to your business, then finite loading might be of great interest.

Level of Detail

One consideration that is becoming more visible is the level at which traditional MRP II plans activities. In most systems, the MRP plan is developed to the day, that is, the system plans a start date and an end date for each activity (purchase order, manufacturing order, or rate-based schedule). This limit of granularity is also common in shop scheduling and priority systems (shop activity control). No attempt is made to schedule within a day. This might have been good enough when virtually all manufacturing was either discontinuous or required days of changeover for continuous facilities, and days of wait time between operations was common. In a flow environment or just one in which activities take minutes instead of days, this limitation could leave quite a lot of manual manipulation to be done to complete the plan. Some newer systems or enhanced versions of older ones now perform their schedule calculations to a decimal fraction of a day (typically one-tenth) which helps somewhat when specifying lead times, but schedules are still developed to the nearest day.

Sequencing of work can also be an important consideration. Even if scheduling is only to the nearest day, in some industries the sequence of material through a machine can be critical. An injection molding operation that deals with colored resins, for example, is easier to run in a lighter-to-darker sequence because little or no cleanup is required be-

From \ To	A	B	C	D
A	×	1.0	1.5	2.0
B	3.0	×	1.0	1.5
C	4.5	3.0	×	1.0
D	6.0	4.5	3.0	×

Fig. 3-3. Changeover time grid used for sequencing.

tween colors. To go from black to light yellow, on the other hand, might require extensive disassembly, cleanout, and reassembly. There are sequencing packages available either as an integral part of the scheduling system, as packaged enhancements, or as supplemental systems that accept the daily schedule from the planning system and apply the sequencing logic to develop the detailed plan. Sequence information is usually presented to the system as a grid indicating the changeover time required between any two combinations of items as illustrated in Figure 3-3.

In a related consideration, facilities are typically defined as a certain number of resources available for a certain number of hours per shift or day. Often, like or similar machines are grouped together under a single facility (work center) definition and scheduled as a group. The foreman assigns the work to the individual machines on a day-by-day basis. Similar to the sequencing function outlined above, enhancement or supplemental software packages (often running on PC networks, no matter what kind of hardware the planning system is on) can be used to sort out the work to individual machines on a finite-load basis, assisting the foreman in this task. An add-on system that does this will usually accept reports of activity to keep the assignments up to date. An integrated system would require that this activity information be passed to the planning system and not force the user to enter reported activity data to both systems.

Just-In-Time (JIT)

Much has been written about JIT, but it is still one of the most misunderstood concepts in manufacturing management. I think the name itself is at least partially responsible for the confusion. The words "Just-In-Time" conjure up the image of a supplier's truck backing up to the side door of a plant, and the delivery person handing a fender to an assembly worker just as the chassis rolls by that position of the assembly line. While this kind of delivery, timed exactly to the need, is at least theoretically possible, it is not at all what JIT is about.

JIT is a philosophy, more than anything else, and does not inherently include any particular software or operating procedure. The real essence of JIT is a focus on waste and a process of continuous improvement.

The JIT definition of waste is "anything that adds cost without adding value to the product." That's an interesting definition because it touches all areas of the business. To start with the obvious, inventory certainly adds cost but its existence doesn't add to the product in any way. Therefore, inventory should be eliminated, or at least reduced as much as possible. True elimination would result in delivery from the supplier (or another portion of the plant) exactly at the time of need.

Seldom is it possible to time things quite that perfectly; and even if you could, you would probably be reluctant to run quite that close to the edge of disaster. After all, if something should go wrong, you'd be out of business until the problem could be corrected. But that doesn't stop people from applying the idea behind JIT to reduce inventory as much as possible, consistent with the realities of business.

Inventory can usually be greatly reduced through the effective application of management systems like MRP which help identify the needs and the opportunities. The only inventory needed in a properly managed MRP implementation is enough to buffer against any uncertainties such as possible late deliveries, quality problems, errors in inventory or bill-of-material records, etc., and whatever inventory that might result from lot sizing considerations (price breaks, minimum quantities, etc.).

The difference between JIT and MRP is that *MRP is a defined set of techniques* (calculations) that helps us manage given the current situation. *JIT is a philosophy of improvement* which might use tools such as MRP to assist in the identification and management of problems, errors, and opportunities for improvement.

Another wasteful item in most plants is handling and movement of work and materials. If a plant can be rearranged to avoid unnecessary handling, such as in a cell arrangement, costs will be reduced with no adverse effect on the product. In fact, quality will probably improve because of the specialization and the focus that cellular manufacturing engenders.

Speaking of quality, lack thereof can be quite costly as in inspections, rework, scrap, and unhappy customers. Complete control of the process, such that no bad parts are produced, using Statistical Process Control, and by enlisting employee involvement, through aggressively implemented suggestion programs and quality circles, can result in a lower production cost, shorter lead time, and an improved image in the marketplace.

Traditional MRP II is not concerned with the measurement of quality. Of course, quality has an impact on schedules, costs, activities, and priorities, which certainly impact the accomplishment of the plan. But remember that MRP II is a set of tools to help you deal with reality, not necessarily a mechanism aimed at improving the situation. Part of that reality is a statement of the expected impact of any quality problems. If scrap, rejects, and rework are expected, there should be an allowance for them in the definitions or buffers used by the planning and execution systems. If quality improves, the definitions should be changed to reflect the new realities.

Nowhere does MRP II have any need for parametric measurements (see potency, yield, and grading discussions in Chapter 7). Statistical Quality Control (SQC) and Statistical Process Control (SPC), although important parts of modern manufacturing management, are relatively separate subsystems, either add-on's to the MRP II system or residing separately on other hardware, often on a plant-floor PC network.

You will see the term JIT applied to MRP II software offerings as a feature or function of the package. What the vendor is usually selling is a form of synchronization between feeder processes and/or electronic replenishment (KANBAN) facilities. Both of these features are discussed in Chapter 4.

Summary

Shop floor control for discontinuous production is based on scheduling and priority subsystems for work orders. As each order progresses through the shop, priorities are generated that indicate how early or late the job is relative to its MRP-generated due date. It is assumed that each job will spend a certain amount of time waiting for its turn in "queue" at each work center that it visits. Management of the flow of work uses resequencing of waiting orders in the queues to either accelerate or retard their flow into the facility.

This kind of work flow management assumes that there is a balance between input and output. If more work is introduced into the shop, or any facility within the shop, than can be completed in the same time period, average queue time will increase as will lead time for all future work flowing through the facility. If less work is introduced than is completed, average queue and lead time will decrease. The actual wait time, and therefore lead time, for any particular job will probably depend more on how it is sequenced at each operation than the actual amount of "active" work time.

Shop floor control systems in different MRP II packages differ mostly in the details: the scheduling calculations used, format of reports and inquiry screens, details of data entry, and editing processes. The most significant difference found is the option of scheduling work according to a finite limit of resource capacity versus the more common "infinite" loading assumption. Either approach is viable, and the choice is mostly personal preference in conjunction with the specific needs and situation of the plant.

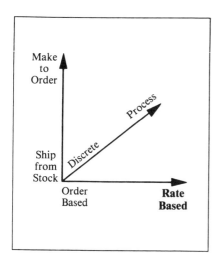

4. Continuous "Flow" Production

Not too long ago, flow production was used primarily in two situations: highly automated, high-speed high-volume production; and in a "transfer line" area, usually for final assembly and often with many (human) arms and legs surrounding the "line." Today, the flow concept is being applied much more liberally and in a much wider range of circumstances.

One of the results of waste reduction efforts (the JIT philosophy) is a heightened recognition that unnecessary handling adds cost and lead time to production and therefore should be minimized. At the same time, reducing the manufacturing cycle (time) has become a major competitive weapon in many industries. The traditional plant layout—grouping similar facilities together—tends to increase handling and the "mileage" that parts travel within the plant as they move from one operation to the next, bouncing around the plant in their circuitous path from department to department.

A simple solution to this situation is to rearrange the facilities so that a series of operations can be performed on a part with little or no handling in between. This requires that dissimilar facilities (machines) be grouped together and that work is assigned to such a grouping based on the arrangement and capabilities of the group or cell.

Within the cell, parts flow (sometimes) continuously from one facility (machine) to the next. From a schedule and control viewpoint, the cell can be managed as if it were one facility. There is little justification for reporting activity within the cell, only at the output end.

The cell concept is very flexible. A cell can be anything from two operations to hundreds. The more complex the production activities in the cell, however, the more likely it is that there will be a more limited variety of parts or products that can be produced therein. When assigning work to the cell, one must consider the nature of the work to be

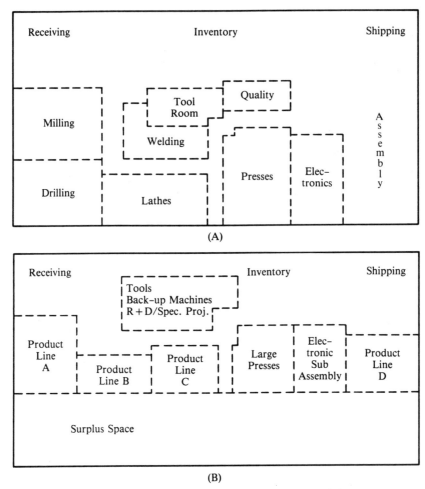

Fig. 4-1. (A) Typical discontinuous production facility. (B) Typical continuous or flow production facility.

accomplished, just as in traditional one-step work centers; but because more than one activity is accomplished, usually fewer different items can be made there.

Most companies end up with a variety of cells, each with capabilities that differ from the others. Parts are assigned based on the match between the cell's capabilities and the processes to be accomplished. To be effective in the use of cells, the design engineers must know what the capabilities of the various production cells are and try to design parts that can be made efficiently in existing cells (or new cells could be developed for the new parts if justifiable).

This awareness, in the design department, of the limitations and capabilities of the production area is a part of what is known variously as

Design for Manufacturing (DFM), Simultaneous Engineering, Concurrent Engineering, or Early Manufacturing Involvement (EMI). DFM is a common consideration in a CIM implementation effort because it not only yields better designs, from a manufacturability standpoint, but can also greatly reduce the design-to-market lead time by overlapping the design and manufacturing engineering lead times. It also helps avoid manufacturing difficulties and redesigns required to correct them.

The chief characteristic of continuous or flow manufacturing (often showing up in the trade press as Continuous Flow Manufacturing or "CFM") is that each unit, when completed at one facility, is free to move to the next step in the process without waiting for others in its group. The application of this concept can have a dramatic effect on manufacturing cycle time.

Flow production is often equated with process manufacturing, but there is not necessarily a connection. Many companies using flow manufacturing are of the traditional "discrete" variety, and there are many process companies that are a batch-mix situation, definitely not flow.

You will sometimes hear people talk about flow production and JIT as if they are the same thing. While many times CFM is one solution brought about by a JIT program, they are not necessarily connected. This misunderstanding is compounded by the fact that many people think of JIT exclusively in terms of KANBAN "pull" signals, and material movement considerations. As explained earlier, JIT is a general improvement philosophy not necessarily tied to any particular management techniques.

In the continuous production world, however, the term JIT is often used to mean electronic signals within the control system for inventory movement (to the point of need at the time of need), whether supplied from stock or from another production facility as in synchronous production. Equivalent terms include "line setting" for synchronized feeder lines and "electronic KANBAN" for the pull signals.

Flow Production and Lead Time

Many studies have shown that, in traditional discontinuous production, nonactive time, primarily queue or wait time, typically makes up 80% or more of the total production lead time. On a job with a total lead time of ten days, the parts can be expected to be just sitting around waiting for eight of those days. Further, if the production quantity for an item is 100 units, it can be assumed that, of the remaining 20% of lead time, only $1/100$ (or less, when you consider set-up time) is active for any given unit. While the entire lot is sitting idle for eight days, each piece is also sitting idle for 99% of the remaining two days. Ideally, such a part could be produced in minutes on a continuous flow line. The actual

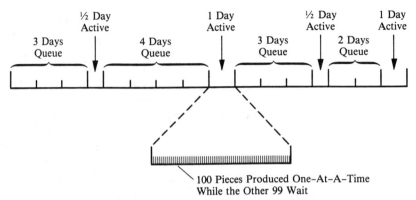

Fig. 4-2. Components of lead time.

production (flow) time for a given unit could realistically be little more than the sum of the time-per-unit for the operations to be performed: no wait, no waste.

Flow Production Management

Traditional MRP II shop activity control applications, designed for discontinuous production, have difficulty adapting to a flow situation. Scheduling assumes that there will be an identified lot quantity of the item being produced, that the entire lot will move as a unit through the process, and that there will usually be some waiting time between operations. While any of these factors can be overridden selectively, the base assumptions are not valid in continuous flow: each part moves through the process at its own pace, there is little or no waiting between steps, and often the need to establish and report against an "order" identification for the lot is a burden that is not justifiable.

In addition, the reporting systems designed for discontinuous production do not lend themselves to the needs of a flow facility.

* Flow manufacturing is not necessarily concerned with costs-per-lot, which is the primary focus of discontinuous systems. They often prefer departmental costs.

* Scheduling by facility, by order, by day is inappropriate. Flow facilities are scheduled by quantity of product on a "line" by day.

* A line is usually a collection of dissimilar facilities. Work centers are usually groups of similar machines.

* Order-based systems assume detailed reporting by order. Flow plants often don't use work orders at all. They want to streamline the

reporting requirements because things happen too quickly and detail is not needed in many cases.

Enhancements to standard MRP can recognize and address the above differences. While some are relatively easy to satisfy, more flexibility in facility definition, for example, others such as scheduling by line are more complex and sometimes are addressed by a separate "module" specifically designed for flow production management that can be implemented either as a replacement for traditional shop-floor controls or as a supplement for that portion of the facility that is flow oriented.

Alternative Shop-Floor Control

The primary issue addressed here is scheduling and managing work flow. In the traditional scheduling process, there are four elements of lead time: move, queue, set-up, and run times. The job is scheduled through each process step by adding the four time factors together, assuming that the entire lot will move through the facility together—all 100 pieces (for example) will be in place at the milling machine work center until all 100 are completed, then they will move together and wait together at the next step. Priorities, dispatch lists, and sequencing decisions are used to monitor and manage flow.

The sequencing of jobs in the queue, more than anything else, determines the actual wait time and therefore the actual lead time for any given job (assuming adequate balance between load and capacity). Presumably, jobs that are late will be processed ahead of jobs that are early and therefore will have shorter wait times at the work centers and shorter overall actual lead times.

In a flow process, the lead time (flow time) for a particular part is determined by the run time at each process step plus whatever movement and wait time there is between steps, which should be minimal. Lead time (flow time) then is the total time it takes for the part to travel from the beginning of the process to the end. It is desirable to synchronize the production rates at each facility on the line such that work will move smoothly through the process with no accumulation of work ahead of a slower facility (or behind a faster one), nor a shortage of work downstream from slower upstream facilities. The overall production rate of the line will be approximately equal to that of the slowest resource. Line speed is normally specified in pieces (or feet or pounds or whatever) per minute or hour. Cycle time for a line is the interval between the introduction of subsequent parts or, alternatively, the interval between the completion of subsequent parts.

The main factors in scheduling a line, in addition to its speed, are changeover time and whether or not there can be overlap. If no overlap is allowed, the changeover time starts after all pieces of the previous item are complete. After changeover is accomplished, there is an additional delay (one flow period) before the first piece of the second job makes it to the end of the line. With overlap, changeover of machines or introduction of new product at the beginning of the line can be started while the last of the previous product is still moving down the path. Changeover is usually minimized with overlap. Changeover and overlap often differ with the products being produced. With a colored product, for example, it may be possible to go from white to yellow to orange to red to brown to black with little or no cleanup or runout in between. To go from black to white, however, may require a complete teardown and cleaning. Scheduling processes should recognize these differences and allow the user to vary the changeover parameter as the activities are loaded and sequenced.

The scheduling process designed for discontinuous production is usually too crude to be applied to a flow process. The time scales of flow production usually involve products that are produced in large quantities during a defined period (minutes, hours, or sometimes several shifts duration) with components required and product becoming available throughout the time span. Ordinary production control systems would typically schedule only to the nearest day, and would want to space out the activities over sequential days rather than completely overlap (contiguous and concurrent). Finer granularity is needed (increments of less than one day), or no detailed scheduling or tracking is needed at all. In many cases, traditional MRP is used to develop the overall schedules, and either manual procedures or off-line facilities—PC spreadsheets, for example—are used to sequence and track production within each day or shift.

Several widely used MRP II systems (including several of the most widely installed systems: IBM's MAPICS/DB and BPCS by System Software Associates) now include repetitive or continuous production management capabilities either as a separate "module," as is the case with MAPICS, or integral to the product and process definition (BPCS).

In the IBM product, the MRP planning logic has a special smoothing technique which is applied to "schedule controlled" items so that planned production can be leveled more than is usually seen in order-based production. The planned daily schedules are released to the repetitive production tracking module where they must be sequenced by the user since there is no logic in the system to recognize such things as color changes.

Scheduling, tracking, and reporting in this system is by item, by line, by day. No traditional work orders are generated, although the system tracks the data internally using the same kind of file structure as it uses for orders. In this case, instead of an order number, there is an internal schedule identification which ties to an item at a given production quantity on a line on a particular day. While the user can see this schedule number on some of the displays and reports, it is not a key to accessing or reporting information. The system also assumes that there will be little or no reporting of activity other than the quantity produced, although intermediate reporting points can be defined. All production data are back-flushed from the quantity produced, i.e., earned labor and overhead, and use of materials.

In the BPCS product, a special "order policy code" identifies make-to-schedule (repetitive) items which can be scheduled, sequenced, and produced on defined flow facilities. As in the IBM product (which has a code among the item characteristics identifying "repetitive" items), reporting is primarily done on completion of the product with theoretical (back-flushed) labor, facility, and material usage generated at completion.

KANBAN Material Control

As mentioned earlier, despite the confusion over terminology (JIT, KANBAN, synchronous production, line setting, pull signals), many systems now provide a capability to synchronize the movement of materials to the continuous flow line based on the schedule and assumed usage during the day or shift. I'll use the term KANBAN.

KANBAN is a Japanese word meaning card, derived primarily from a system developed by Toyota to time the production and movement of common component parts using paper cards or other devices as a physical replenishment signal. The term now means the signaling of replenishment of materials to the production floor by means of a physical or electronic signal tied to the usage or need of parts by the "using" production facility.

Let's assume, for example, that an assembly line is set up to produce 1000 units per day of a variety of metal boxes, and all of the boxes use the same hinges which are produced on site in another production area. At the beginning of the day, a box of 100 hinges is brought to a convenient location in the assembly area. There is a tag attached to each hinge. The assembly worker is instructed to take the hinges from this box as needed, but must place the tag from each hinge used on a hook above the box.

An inventory control person can now circulate through the assembly area and, when he or she passes the hinge station, collect the tags from the hook. His or her job is to attach these tags to new hinges from stock and put them in the box. In this way, there will always be a supply of hinges ready for the assembly person as long as the inventory person keeps up.

Another way that this can be done is to use, let's say, three boxes or trays of 50 hinges each. As the inventory person observes the assembly area, he or she will look for empty trays. The container is now the KANBAN or physical replenishment signal. The KANBAN, whether tag or tray, replaces the traditional "pick list" that controls replenishment.

Extending this example one step further, remember that the production facility for hinges is on site. Instead of timing the issue of new hinges from stock, the tags or empty trays can be used to trigger production of more hinges. The tag now replaces the traditional work order (it is the authorization for the hinge line to make replacements) as well as a material transfer ticket (pick list), and the finished hinges can move directly to the point of use—Just-In-Time—or pretty close to it.

The production rate for the supplier line or area, sometimes referred to as a feeder line, is now synchronized to the rate of use of the assembly line. The number of hinges in stock and in process is nearly equal to the number of cards (KANBAN's) in circulation (less any tags that are still on the hook). To increase the supply or production rate, increase the number of KANBAN's. To decrease the supply, take cards or bins out of circulation.

KANBAN can be a very effective technique in circumstances similar to the example. It is simple, effective, and easy to implement. The biggest challenge is to coordinate the KANBAN process to the master schedule, making sure that the proper number of the correct cards (items) are in process at the right time.

As explained earlier, KANBAN is really a refinement of an early inventory control technique called "two bin order point." Because it is in the order point family, replenishment occurs because of past usage not future need. That is why the technique must be coordinated to the master schedule to be effective.

Here's what can happen with KANBAN. If the product(s) that uses the hinge is only being produced for a part of the day, and the line will be switched over to another product line at noon, you are likely to end up with a large supply of hinges left over when the line is stopped for changeover. Replenishment continues until all of the KANBAN's are attached to completed hinges. Manual controls are needed to stop replenishment when enough hinges are available to finish out the production run of boxes, or perhaps when there are enough hinges left over to "prime" the next production run.

The MAPICS/DB continuous production module mentioned earlier incorporates what is referred to as an *electronic* KANBAN system that transfers inventory from a defined supply location to "line" locations in response to the released schedules. Only enough inventory is transferred to complete the requirement for the day or to fill the location, whichever is less. Replenishment is triggered by the back-flushed usage, subject to the same limits. This version of KANBAN is tied directly to the master schedule through the MRP process, and the quantity of components moved to the line will not exceed the quantity required.

It is also interesting to note that in the MAPICS system a component or material can be identified as a "JIT" item. In this case, the component is not issued (moved) to the line but is expected to be there when needed. If the component has not been moved to the line by the time production is complete, a warning is issued that interrupts the back-flushing process. In this implementation, the term JIT refers to components or materials whose movement is not tracked by the system, therefore presumably controlled through physical KANBAN's or some other method.

Mixed-Model Scheduling

Scheduling of production on a continuous flow facility can be done in two different ways. The foregoing discussion alluded to a schedule to produce some quantity of an item, then changeover to another item. This method is called "campaign" scheduling, and it assumes that all of the items produced in a "run" are alike. The alternative is called "mixed-model," and it allows products to be intermixed on the line. Obviously, mixed-model scheduling requires that the line be capable of producing all of the items scheduled with little or no changeover in between.

Mixed-model scheduling offers the ultimate in flexibility. Production can be timed to the exact needs of the customer or subsequent facility, and any quantity of any product (within the range of products produced on the line and within the total production capability) can be made as needed. Often, a mixed-model line is equipped with intelligent controllers that can either receive schedule (sequence) information from the planning and control system or read a label on the workpiece or its holder that identifies what model is to be made.

Planning and scheduling for mixed-model is a challenge for traditional MRP systems. Since there is no "lot quantity" in the traditional sense, normal tracking and control systems don't work. In a specialized "repetitive" or flow control system, the normal facilities can be used because scheduling is done by line, by day. If the schedule calls for 100 each of products A, B, C, and D, these can be "loaded" to the line and reported as each unit completes. As long as there is no changeover to be considered, reporting of an A, two B's, another A, then a C, etc., is really not much different from reporting 100 A's, then 100 C's, etc.

Campaign: AAAAAA BBBBBB CCCCCC DDDDDD EEEEEE

Mixed Model: EABDCEACBDDEBABCCADBEEACDCBDAE

Fig. 4-3. Campaign and mixed-model scheduling.

Any capability or interest in sequencing of products on the line must be handled differently in mixed-model scheduling. Several approaches are possible including the following.

* *Automatic sequencing by priority.* I don't know how much software there is to do this, but I suspect that the results might be disappointing. Priority for a group of items with the same routing and lead time (as they must have if they run on the same line) can only be by due date in most MRP systems, which is a rather crude measurement. All items with the same due date must run on the same day; so what triggers the sequencing decision?

* *Spread evenly.* This is probably the default solution. An MRP system can plan a needed quantity for a certain date interval, and the quantity can be spread evenly (smoothed) so that the same quantity of an item is produced each day. This would yield the most level schedule. This technique does not address the sequence within each day—all the A's first, then all the B's, etc., or two A's, then three C's, another A. . . .

* *Pattern.* The user could supply a desired pattern and the software could attempt to duplicate the pattern or apply rules for sequencing using an artificial intelligence approach.

* *Manual.* The schedule requirements can be presented to the user unsequenced, and facilities can be provided to manually designate the run sequence to be used.

Capacity

Line capacity is of primary concern in most flow environments. While finite scheduling logic can be applied to the repetitive situation, a more typical approach is to establish and recognize line capacity on the master scheduling process, then build in a capacity check at the schedule release stage.

Resource Requirements Planning (RRP) and Rough-Cut Capacity Planning (RCCP), which are both master scheduling tools for checking the validity of the plan, are typically set up to recognize a capacity in terms of units per day (or week or month), and to test the plan (master schedule) against the identified capacity. Both RRP and RCCP are designed to work only with identified constraining or "critical" resources. Fortunately, in a production line, all resources are coordinated together to run at the same rate, a rate usually stated in terms of pieces per hour or its reciprocal (time per piece or cycle time), which is easily converted to a rate (capacity) per day for the line.

Further, a production planning family of items can be defined to include all items that run on a given line, thereby making RRP, which is tied to production planning, coordinated with the need to recognize the line load/capacity relationship. In this way, the master schedule can be developed with a nearly direct tie-in to line capacity limits.

When specific schedules are released in the execution module, most systems provide feedback to the scheduler that shows hours of line capacity available and hours consumed by the schedules being released. This is, in effect, a form of finite loading because any overload is immediately flagged for resolution.

Back-Flushing

One of the assumptions mentioned several times in this chapter is that of back-flushing. While used in traditional discontinuous situations as well as in flow facilities, back-flushing is more common in a continuous environment because of the desire to minimize reporting and less interest in tracking detailed status within a line or lot of product.

The concept of back-flushing is simplicity itself. When you report completion of a quantity of a product, usage of materials (and perhaps other resources) is assumed based on the quantity complete. Back-flushing uses the standard bill-of-material to determine the materials used and the standard routing to calculate the labor and overhead.

As in all other things, when you take a shortcut, there is a price to pay. With back-flushing, what you gain in convenience and reduced reporting you lose in visibility and accuracy. Since standard definitions are used, no variances will be detected unless outside reporting and reconciliation are applied for such things as scrap, nonconforming materials, and labor variances. The reduced visibility is a result of the fact that there is no intermediate reporting so you cannot track progress through the facility—you only know what is completed.

These limitations are usually not of great concern to the continuous manufacturer. A bigger problem occurs when back-flushing is applied to a noncontinuous process. Many companies back-flush materials in order to avoid having to report material issues to production. The lack of "actual" data seriously compromises the validity of the production reporting system, the actual costing system, and inventory accuracy.

Departmental Costing

With no clear identification of a "job" or work order in a continuous environment, the traditional job costing system in most MRP II systems no longer applies. Rather, flow manufacturers often accumulate cost information by area of the plant by time period, and apply these costs to the products produced during that interval. A department can be identified that produces a family of products, and costs associated with that department such as direct labor, indirect labor, quality management services, utility costs, building and maintenance allocation, administrative allocations, etc., are calculated for a period. The quantity of product produced during that period is determined and divided into the total cost. The result is the cost per unit for the products in that group.

This approach necessarily assumes that all products within the group use the same amount of resources since there is no tracking of the actual resources used (back-flushing). In many cases, this is a reasonable assumption. A packager of liquid product, for example, might calculate a total cost of $500,000 per month for labor and overhead for a production facility. If 2 million units are produced in a month, the calculated cost per unit is 25 cents. It probably doesn't matter that the products include half-pints, pints, quarts, half-gallons, gallons, and liter bottles. As long as the line speed is similar for all sizes, it is not unfair to apply the same cost to each size.

This general approach is called departmental costing if the rate is determined for a department, operation costing or line costing if the basis of measurement is an operation or a line, or sometimes process costing which can be applied to an entire facility or some other breakdown.

Implementing Flow Manufacturing

The benefits of changing a traditional production shop to flow manufacturing can be dramatic. Reports of an 80% decrease in cycle time (total time-to-produce), 60% reductions or more in WIP inventory, and decreases in scrap, rework, space requirements, overtime, etc., are not uncommon. It is easy to get carried away with the potential and not look at the concept realistically to see what it takes to make it work.

Flow manufacturing, either for the whole shop or just a portion of it, assumes that the products that will be produced in the flow facility require the same processes in the same sequence. There is room for some small amount of variability, but not much. The whole idea is to keep things moving evenly through the process.

Rearrangement of machines is often the best way to reduce handling between process steps but is not strictly required. Relocation of equipment may not be possible because of physical constraints such as the size and weight of the machines or availability of utilities. If there is still travel involved between flow resources, the controls and facilities must be set up to move the materials efficiently. Many flow arrangements use material handling equipment such as conveyors, chutes, or carousels. With physically separated machines, bins, tubs, pallets, and other containers are often used and the material flows not in individual units but in small container-sized lots.

The biggest difference, from a material control standpoint, between batch and flow processing is the existence of inventory buffers between operations. To implement CFM, eliminate these buffers (and adjust the planning/controls and measurements that go along with them). It is not wise to completely eliminate these buffers all at once. As you reduce the buffers (lot size), it is likely that problems will surface that were hidden by the inventory and time delays between steps.

Two major considerations are quality and equipment reliability. Quality problems, in batch production, can be compensated for by increased production. If a 10% scrap rate is expected, the lot size can easily be increased by 10% and the nonconforming parts discarded when they are detected. With continuous flow, production rates are synchronized. When a bad part is produced at step four, all downstream facilities will have to wait until a replacement is produced. The entire production process, and rate, is affected by defects.

The same effect happens with machine problems. If one machine stops for any reason, all downstream facilities will be starved until the machine is back on line. When implementing flow manufacturing, you must study the reliability of all facilities, as well as their production rates, to plan for maintenance and repairs. A solution, when reliability cannot be assured, is to have a centrally located back-up facility available. If there are a number of cells in your plant, have extra machines of each type located in a central area that can be used to keep production going while a machine is being repaired. The material flow will be somewhat disrupted when these back-up machines are in use, but production can proceed when it would otherwise be completely shut down. These machines could be used for prototyping or other custom production when not serving in their back-up role.

Quality Impact

Since quality is so much more important with synchronized production, companies implementing CFM necessarily take a greater interest in process control and quality improvement. Fortunately, the flow environment helps in the identification and resolution of quality problems.

With batch production, quality problems sometimes are not discovered until days after they happen. If operation three produces bad parts on Monday, they might be sitting in the queue at operation four until Thursday. Finding and correcting the source of the problem three days later is difficult if not impossible. With flow production, you will usually know about quality problems within minutes. Getting to the source of the problem and, by the way, preventing the unknowing production of more bad parts, happens immediately.

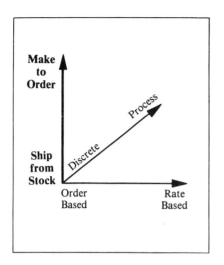

5. Make-to-Order, Ship-from-Stock, and In-Between

This dimension of our variations matrix refers directly to the finished goods inventory policy and is primarily concerned with the requirements and parameters of the master scheduling function. There are additional considerations, however, for custom manufacturers who respond to customer demands on very short notice with nonstandard products. The considerations for quick response, custom product, and the additional requirements of government contracting are covered in Chapter 6.

The primary determinant of master schedule development is the relationship between the total lead time that it takes to produce the product—known as Cumulative Material Lead Time (CMLT)—and the span of time between the receipt of the customer order and shipment of the merchandise. If the promise date is beyond the CMLT, this is truly make-to-order. If the item is completed and in finished goods inventory before the order is received, that's ship-from-stock. Most companies and products fall somewhere in between these two extremes.

You might hear some people talk about the D:P ratio. This is another way to express the relationships just presented. In this case, D stands for delivery time or lead time promised to the customer; P stands for production lead time, which could be either CMLT or only the production portion of total lead time. If the D:P ratio is greater than 1, you can make-to-order. If D:P is less than 1, you must begin your activities before receipt of the customer order. Ship-from-stock is a D:P of zero.

Cumulative Material Lead Time

CMLT is the total elapsed time to produce an item and, as illustrated in Figure 5-1, includes all production and purchasing activities, arranged

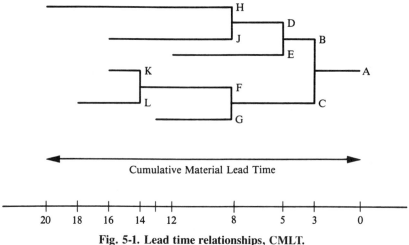

Fig. 5-1. Lead time relationships, CMLT.

as in a critical path analysis. In theory, if a customer order is received today, and there is nothing started and no materials in stock, the minimum time in which the item could be produced, with no expediting, is CMLT (twenty days in this example).

Of course, it is possible to reduce this total lead time in a number of ways. One easy way is to stock low-level components. In this illustration, maintaining a supply of component H immediately reduces the overall lead time to eighteen days. Stocking H plus L brings lead time down to sixteen days. Adding K and J to stock, along with H and L, reduces the total lead time to fourteen days. At this point, the longest leg is made up completely of manufacturing activities. The common name for this range of time is Cumulative Manufacturing Lead Time (CMfgLT).

Notice that the elimination of item H as a concern changed the critical path from A–B–D–H to A–C–F–L. If your goal is to reduce CMLT, you must be aware of how the critical path moves, in order to be able to identify lead times that directly contribute to CMLT. You could easily spend time, effort, and dollars chasing lead times that don't matter, i.e., those not on the critical path.

Other ways to reduce CMLT include reducing individual lead times for critical items. Instead of stocking the lower level components, it may be possible to forecast their use and place them on order in anticipation of the need. In the first example above, instead of stocking item H, it is only really necessary to anticipate the customer order by two days and initiate the purchase order for H two days before the order is actually booked. As long as H comes in on time, that is, eight days after the customer order is received, the eighteen-day total lead time can be met. Working with your vendor to reduce lead time or finding a vendor who can deliver sooner is another option.

Reducing the manufacturing lead time for an item can also have a similar impact, as long as the manufacturing lead time is on the critical path. Reducing the manufacturing lead time for item A by two days reduces the CMLT by two days, no matter what else happens, because final assembly is on everyone's critical path.

Why reduce lead time? It's probably obvious that quicker delivery to your customers is a competitive advantage in most instances. In fact, it is widely agreed that today lead time is the most critical success factor in many major markets. In a world-competitive marketplace, quality and price are assumed: if you can't meet your competition in these two areas, you can't even play the game. Being able to bring a new product to market ahead of your competitors is often the key to success and profits. If the product is successful, the others will be there soon, and price competition takes over until the next innovation starts the cycle again.

Similarly, those manufacturers that are most flexible, that are best able to change their production schedules on short notice to track changes in demand, will have less obsolescence, less inventory of all types, and better customer service accomplishment. This applies, obviously, to ship-from-stock, but is also important for make-to-order. Quoting a shorter lead time to the customer will often get the business, as long as you can deliver.

The situation for the ship-from-stock company is the opposite of make-to-order, on the planning scale. From-stock products are produced to a forecast and must be on hand when the order is received. Lead time is still important, though, because shorter lead times provide greater flexibility. In order to be able to plan all activities—acquisition of purchased parts, fabrication, assembly, and final assembly—you must forecast at least one Cumulative Material Lead Time (CMLT) into the future. If CMLT is reduced, you will have more time to refine your forecast before you will have to commit to a schedule. With a shorter forecast horizon, you should be able to forecast better. It's a lot like throwing darts at a target—you should be able to hit more bull's-eyes if you are standing closer to the target.

Of course, things change. And the one thing that everyone knows about a forecast is that it will not be 100% right. The more accurate the forecast, the better your customer service will be. One way to compensate for forecast error is with an inventory buffer (safety stock). Better forecasts support a higher service level with a given buffer investment, or allow a lower investment for the same service level. The three factors are tied directly together, and you cannot change one without changing at least one of the others. See Figure 5-2.

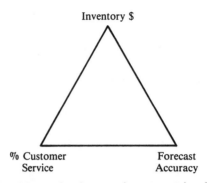

Fig. 5-2. service–forecast–inventory triangle.

Companies that don't understand this are destined to travel back and forth on the same frustrating path: inventory is increased in an effort to improve service level (which will usually happen) until accounting declares that inventory is too high. When inventory is reduced, service deteriorates until complaints and lost business become unacceptable—inventory is then increased to improve service until...

Most products fall in between the two above situations—promised delivery is longer than immediate but shorter than CMLT. As in the first examples, lower level items can be stocked, usually based on a forecast (no matter how informal), and/or purchasing and production activities can be started in anticipation of the order, again to a forecast. When the order is received, the order rate is compared to the forecast, and adjustments might be made if they are not too disruptive or expensive. Here is an area where many companies get into trouble.

Unfortunately, the sales and marketing organization is usually not aware of the importance of CMLT and the costs of changes to the master schedule within CMLT. They have been highly motivated to secure orders and have learned through experience that almost anything goes—manufacturing, more often than not, performs miracles on a routine basis and satisfies unreasonable demands by meeting too-short delivery schedules. In so doing, they have taught sales that there are no restrictions. "You sell it and we'll make it."

The further you get into CMLT (closer to product shipment), the more disruptive, more expensive, and less possible it becomes to comply with the changes or prepare for them with inventory or anticipatory activities. It can be done, but at what cost?

Master Schedule Planning

Master scheduling consists of two distinct processes: one is strictly a calculation technique, the other involves the application of judgment. The three stages of master scheduling are the demand statement, the finished goods inventory plan (requirements), and the build plan (master schedule).

DEMAND
judgement
REQUIREMENTS
calculations
MASTER SCHEDULE

The demand statement can include a forecast, a backlog of customer demand (orders with future ship dates), manual input, or a combination of these. The finished goods inventory plan is a statement of what quantities of each item you would like to have available to ship, by date. This plan, called statement of requirements, is the target that the planning system is designed to address. The resulting master schedule is the calculated best way (given the conditions and instructions in the planning system, such as inventory availability and lot sizing rules) to satisfy those requirements through manufacturing and purchasing activities, that is, how to avoid any projected shortages with minimum excess resources.

The calculation process between the requirements and the master schedule is really quite simple and consists of three of the four steps of MRP as described in Chapter 2. The missing step is the bill-of-material explosion (gross requirements), since master scheduling is limited to the planning of sellable items and not their components. Master schedule generation begins with the netting of the requirement against available inventory. As in MRP, any net shortage is satisfied first by moving existing firm planned or actual ongoing activity to align with the need, then planning new activity (MRP step three, order planning and MRP step four, lead time offset) to fill in any remaining gaps.

The real challenge in master scheduling is the development of requirements from the demand statement. This part of the process is not mechanical and employs the business judgment of the master scheduler and the company executives.

The demand statement, for example, might identify a market for 10,000 widgets per month for the next 12 months. If your demonstrated capacity is 8,000, it would be foolish to build a master schedule to completely satisfy the demand—a master schedule that is beyond capacity will not be fulfilled, will build inventory (components and work-in-process), and will actually lengthen lead time by overloading the plant. In another example, seasonal demand might exceed capacity during the peak period but not fully utilize available resources off peak. You may want a master schedule that levels production to avoid overtime, layoffs, etc.

In addition to capacity considerations, adjustments between demand and requirements can also include considerations of product mix, cash flow, desired margins, market penetration, and many more strategic considerations. Managing the master schedule in this way translates company strategy into finished goods policy which drives all other activities. Master scheduling and MRP establish the operating plan for the company—how resources are to be utilized to meet corporate objectives.

From Stock

In ship-from-stock situations, it is necessary to have the item manufactured and in finished goods inventory before the customer order is received. Production is based on some kind of forecast, no matter how informal.

Most people don't like to forecast. Forecasting is usually a no-win situation: forecasts are always wrong, and being wrong is no fun. No matter how sophisticated or how accurate the technique, the further ahead you forecast, the less certain and less accurate you can be.

In most cases it is the sales/marketing function that has the best ability to measure the pulse of the marketplace and develop an estimate of expected future demand. Unfortunately, they are often reluctant to do so, or do so only in a very general way, i.e., total dollar sales, dollar sales by product group, or units by product group. In order for the forecast to be useful for production planning, it must be stated in terms of units per individual product or SKU* per date or period.

Since production (planning) needs the forecast and often can't get it from marketing, they end up developing their own, based on production history, which might not accurately reflect sales or demand history.

One characteristic of a forecast of any kind is that it is most accurate in the short term and gets less accurate and less certain further out. Think of it like aiming a dart at a target; from a few feet away, it's not too hard to hit the center. Move the target further away, and accuracy will suffer. Like the dart board results, the forecast will be less accurate the farther it extends into the future.

If your ship-from-stock item CMLT is three months, then in order to have the right products in stock to meet projected demand it is necessary to forecast at least three months ahead and produce according to this forecast. If the CMLT is six months, the commitment is locked into place

---

*SKU = Stock Keeping Unit, a common generic term for finished goods inventory items in wholesale and retail.

based on a forecast that is six months ahead—one that is bound to be less accurate. On the other hand, if CMLT could be reduced to two months, you will now have an extra month, compared to the three-month CMLT example, to refine the forecast before locking the plan in place.

Just to complete the picture, let me state that forecast accuracy can be measured and characterized statistically, and a desired high level of customer service can be achieved by buffering with safety stock (or a safety requirement in MRP) to compensate for the inaccuracy and variation in demand. The better the forecast, the less safety buffer is required to achieve the desired service level (less uncertainty to cover). Another way to look at it is: the same safety investment could buy a higher service level if there is less variation (uncertainty) due to a more accurate forecast. A shorter-range forecast is bound to be more accurate, all else being equal.

Ship-from-stock companies know that they must forecast. Usually, they also realize that there will be some uncertainty in the forecast and that there is a real risk of shortages or of excess at the finished goods level, depending on the actual demand in any given period. These companies can be served by a strong forecasting function, tied to the order entry system, which can develop a statistically derived prediction of expected future demand and pass it to the planning (master scheduling) function.

The forecasting system must allow the user to review the result of its analysis and make adjustments as necessary. A statistical forecast is based on past behavior. If the user has knowledge (or intuition) of future events that may cause the actual demand to differ from past patterns, then this external information is applied on top of the mechanical prediction as an adjustment. A good system will also be self-monitoring, and alert the user to any demand that doesn't fit within the parameters of the statistical model.

Statistical monitoring involves averaging or smoothing. Since actual demand will vary from day to day and period to period, smoothing techniques suppress some of this variation to provide a more stable schedule. Smoothing can hide trends and unusual events, however. A good statistical forecasting package will monitor actual demand, warn the user when things seem to be going wrong, and also adapt to the situation by modifying the smoothing technique to allow the most recent data to more strongly influence the regenerated model; a process called adaptive smoothing.

Make-to-Order

At the other end of the scale, the make-to-order manufacturer has no need to forecast in order to develop production plans because he or she can use order backlog as input to the master scheduling process, as long as CQLT (Customer Quoted Lead Time) is greater than CMLT. If it is desirable to reduce the effective CMLT by stocking lower level components or starting the acquisition of parts before the actual order is received, then a forecast, in conjunction with MRP, can be used to identify what parts and materials are needed, when they are needed, and when acquisition activity should begin.

True custom manufacturing, in the strictest definition of the term, would mean that each product is unique. Since there is no way to know the composition of future products until they are ordered by the customer, there is no way to forecast, plan, and reduce lead time by starting ahead of the receipt of the order. But companies are faced with this challenge all of the time. To be competitive, they often must be able to deliver in considerably less time than full CMLT despite the difficulty in forecasting unknown products.

There are several approaches to address this dilemma. The first is to use a "planning" bill-of-material in conjunction with a general forecast to begin acquisition of materials. A planning bill-of-material is a bill for an item that doesn't really exist. The parent item often represents a family of products with some common characteristics. You might object at this point that common characteristics can't be known because every product is unique. That may be true, but in most cases there are truly some common characteristics that can be identified fairly easily and often quite accurately.

Even though the exact same product is seldom or ever made again, most companies are in a particular market segment that can be identified. A machine shop, for example, that usually works on metal parts, can define a planning bill that links raw material types to an overall level of business. If half of the business involves machining of steel blanks, a third is done on sheet metal, and the rest involves rod, the first level of the planning bill might look like this:

	Machined Parts		
	Steel	Sheet	Rod
Quantity-per	0.500	0.333	0.167

The next level would define the content of each of the subcategories such as (under "rod") half-inch 0.500, three-quarter inch 0.200, and one inch 0.300. The next step is to predict the total demand or production level for the entire group, and let the MRP process translate, through the quantity-per and the lead time between each level, a material plan according to the forecast. There are actually two different forecasts here. One is obviously the overall level of production; the other is the distribution within each group as stated in the quantity-per figures. (See Figure 5-3.)

It is sometimes even easier than that. Although the specific configu-

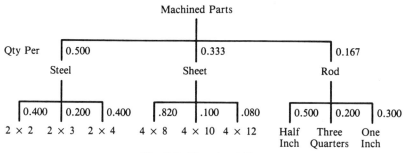

Fig. 5-3. Planning bill.

ration or accessories might differ from one unit to the next, there might be a common basic unit or a number of common parts or assemblies. The normal bill-of-material relationships can be used for these common items, and only the distribution of the different members of each accessory (option) group must be predicted in composing the planning bill. Some systems have a built-in capability to specify a planning bill (or modular bill as it is sometimes called) in this way for forecasting and planning purposes, then allow specification of unique configurations, per the customer order, for assembly from stock parts. This assemble-to-order capability is sometimes called "features and options," "feature selection," or something similar.

When using a planning bill, it is necessary to "consume" the forecast as specific orders are booked to prevent overplanning. In our machine shop example, we would never take an order for a "machined part" item. The customer order would call out a specific item, maybe one never before identified. The materials required to satisfy this order have already been included in the plan for the "machined parts" planning item (assuming that the planning was done effectively), and adding the specific item to the plan will duplicate the need for these materials.

If the actual order is received far enough ahead to cause concern about overplanning (more than one planning cycle before projected completion), the forecast for the "machined parts" item should be reduced so that the total number of planning bill items plus actual order items is equal to the original forecast quantity. If the actual orders exceed the forecast, you must decide whether to change the master schedule to accommodate the demand.

Assemble-to-Order

A variation of the make-to-order environment is assemble-to-order. In this situation, major assemblies or components are built and stocked to a forecast, and the specific configuration of the product sold will differ from one order to the next.

Imagine a manufacturer of bicycles. In one model of bicycle, this manufacturer offers a choice of five frame colors, five wheel styles, five handlebar configurations, five fender options, and five kinds of seats. If each combination were given its own product number and specific bill-of-material and routing, there would be more than 3000 of each to build and maintain. Furthermore, if marketing added a horn option, the number of unique models would immediately double.

An easier way to manage the situation would be to stock frames, wheels, handlebars, fenders, and seats, and assemble the specific configurations only after the customer order is in hand. Assemble-to-order lead times can be very short, and the stocking requirements for 25 major components are a lot less daunting than for 3000 individual models.

Assemble-to-order can be addressed in several ways by MRP II software. One way is to create specific bills as outlined above. If the number of unique combinations is modest, this can be a viable approach. Forecasting for each model can be simplified by combining the models into a group or "family" either through a facility designed for this purpose in the master scheduling system (production planning) or through the creation of a fictitious higher-level item (bicycle family), with the products themselves structured as components under the family "item" (planning bill-of-material). The quantity-per for the products within the family is set to reflect the expected (forecasted) distribution of sales by model within the family, and the forecast is made for sales of the entire family at the fictitious item level, as illustrated in Figure 5-4.

The chief drawbacks of this approach are that there are still specification bills for each member of the family (each specific configuration), and the master schedule must be "consumed" as customer orders come in as they will be entered into the system per the actual product item number, whereas the forecast and master schedule are input according to the fictitious (family) item number.

An alternative is to use a "modular" bill-of-material feature, sometimes called a "features and options" facility, such as the bicycle example in Figure 5-5, in a similar way. With a modular bill, there is one item number for the product (eliminating the need to "consume" the forecast

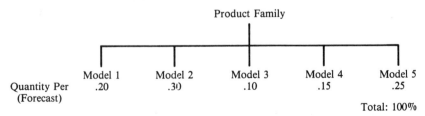

Fig. 5-4. Product family planning bill and forecasting.

MAKE-TO-ORDER

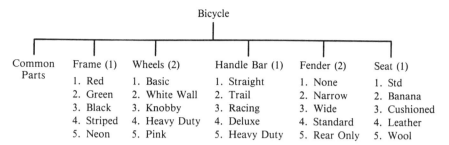

Fig.5-5. Modular bill.

since the item ordered is the same as the item planned), and the choices are structured into its bill definition. No specific bill is stored for a particular configuration. The assembly order is derived from the selections made from the available options.

"Features and options" allows the customer order to specify the choices (options) within each feature that designate the specific configuration ordered. For example, a customer might order a bicycle with options 1, 4, 2, 2, and 3 (a bicycle with a red frame, heavy duty seat, trail-style handle bars, narrow fenders, and cushioned seat). An assembly production order could then be released to assemble the selected options into the desired product. Unfortunately, the item number (simply "bicycle" in this example) does not identify the selected options so there is no way to know, without an outside tracking system, which configurations are in stock—they're all just bicycles. Another challenge is that there is usually only one routing for a given item, and any differences in the assembly (labor and/or resource) requirements between configurations must be ignored or handled outside the main system.

Yet another approach is to ignore configuration and bill-of-material considerations altogether. Using descriptive naming, the customer order and subsequent assembly order can simply describe the configuration. Experienced help can assemble as needed per the description. There are obvious limits to this approach when it comes to planning for inventory and developing cost estimates (standards). This approach is widely used when dimensions are a part of the product configuration such as in the case of window blinds and curtains (many lengths and widths possible), wire cables (an almost infinite number of lengths), and similarly dimensioned products. Systems have a hard time tracking activity, stocked products, required materials, etc., when there is no standard definition of the product's size or content.

If the choices are complex or there are interdependencies to be considered, it is sometimes desirable to use a configurator in conjunction with build- or assemble-to-order specification. Configurators are most often custom designed for the products they will be used with and contain the rules that govern the configuration process.

Manual configurators can consist of books of rules and/or choices such that the user can go from page to page, working through the possibilities (go to the "motor" section, find the chart for 220 volt motors, choose the desired horsepower, find the list of available gear boxes that are compatible with that motor choice, select a compatible control panel, etc.).

Configurator rules and combinations can be built into software facilities so that the order entry operator is presented with a series of interdependent choices, often in a tree-like structure. This ensures that the selected configuration is viable, and it also "walks" the operator or the customer through the choices making a complex process easier by presenting only the acceptable choices and doing so in a logical or convenient sequence.

Configurator software is usually either custom designed or highly tailorable because of the uniqueness of each application. Advanced systems employ artificial intelligence to validate the choices, suggest alternatives, and generally help the operator define the selections. After the selections have been made, the software would generate the specification bill-of-material and sometimes the routings necessary to produce the product.

In-Between

Most companies are neither strictly make-to-order nor ship-from-stock. Some have elements of both, and many others will promise a customer a shipment date that is later than "now" but less than CMLT. As stated earlier, it is quite easy to effectively reduce CMLT* by stocking (the right) low-level components. It is also possible to reduce the lead times of manufacturing processes on the critical path to shorten CMLT. The secret to reducing lead time to the customer is primarily through starting the activities, based on a forecast, before the customer order is received.

If CMLT is three months, and the desired CQLT (Customer-Quoted Lead Time) is two months, simply start the process one month before accepting the order. To reduce CQLT to one month, you must forecast and start two months ahead.

---

*Technically speaking, this is not a reduction of CMLT which is fixed, given the lead times and structure relationships. The ability to quote and meet shorter lead times is due to getting a head start on the process.

The alternative to the above is to fail to recognize CMLT, make whatever delivery commitments you have to in order to make the sale, and let manufacturing do what they can to meet the promised dates. Just because CMLT is a relatively fixed quantity doesn't mean that it is inviolate. Many companies routinely produce and ship products in less than CMLT. They do it with large stocks of low-level parts, large stocks of subassemblies and assemblies, and too much work-in-process. They also do it through expediting, through interrupted production runs to rush through a needed part or two, and through overtime, express freight, and other expensive, inefficient measures.

The biggest problem that many manufacturers face, although it is often difficult for them to recognize the exact problem, is a lack of understanding of CMLT and its importance on the part of the marketing/sales organization. Usually, sales/marketing will accept almost any order, accede to virtually any request, to "get the business." Then, manufacturing must perform miracles to make good on the commitment.

Often, it is manufacturing's continuing success in accomplishing these extraordinary feats that both masks the problem, preventing its recognition and resolution, and cuts heavily into potential profits and efficiency. As one of my clients once put it: "We're just too good at expediting." The company's good record of on-time shipments led management to believe that the low margin problem was due to some fault of manufacturing, whereas the real problem was in making unrealistic customer promises. No one likes to turn away business, and there is little or no incentive to do so if manufacturing proves, over and over, that just about anything is possible.

The only real solution is a combination of education and improved communication. Sales/marketing must learn about CMLT and the effect of changes to the plan within its range.

Any change to the master schedule within CMLT has an impact on some ongoing activity. If the change is up, adding previously unscheduled items or increasing planned quantities, either on-hand inventory, active purchase orders, open production activities, or a combination of these will no longer be adequate to satisfy the actual need unless you happen to have extra quantities on hand or on order. If the change is down, the inventory, or on-order quantities will be too high or not needed at all. Both of these circumstances add unnecessary cost.

The further into the CMLT the change takes place, the more costly the consequences and the less likely it is that the change can be accommodated. Referring back to Figure 5-1, a change in the first two days will only impact the purchase order or on-hand quantity for component H. A change at day eight, however, impacts three purchased items on hand (G, H, and J), one on order (E), and one manufactured part (F). If the change

On-Hand = 0

Date	Planned Receipt	Backlog	Available	Cumulative
3/5	50	50	0	0
3/12	50	40	10	10
3/19	50	35	15	25
3/26	50	35	15	40
-----	----	----CMLT----	----	-----
4/2	50	10	40	80
4/9	50	0	50	130
4/16	50	0	50	180

Fig. 5-6. Available-to-promise display.

is up and there are no extra stocks available, there are twelve days worth of activity to make up as well as accomplishing the remaining eight days of activities, all in the eight days remaining until the promised ship date.

One tool that can be a great help in providing marketing/sales people with some insight into what can be promised and what is not practical is a function of the master scheduling application on many MRP II systems called "Available-to-Promise" or A-T-P. Figure 5-6 illustrates this feature.

The first column of data lists the expected receipts of this finished goods item by date and quantity according to the master schedule. The second column shows the sum of open customer orders by promised ship date. The third column includes the differences between the first two columns, or the net uncommitted expected future production receipts for each period. The last column is a cumulative total of the net available.

If a customer calls in and asks for 100 of these items, and requests a ship date of March 26th, how would you respond? The technically correct answer should be: "I'm sorry, Mr. Customer, but we won't be able to meet that schedule. What I can do for you, however, is ship you 40 on the 26th and the balance in April. Would that be okay?" Unfortunately, a more typical response would be: "How would you like those shipped, by truck or air?"

While this display shows the uncommitted future inventory, it doesn't show to whom the commitments have been made. If the new order is important enough, it is conceivable that you might want to divert some of the planned production away from a previous commitment to the new one. Of course, this still results is a disappointed customer, but those are the choices that you must make. Often, the schedule is changed to expedite more product to meet the increased demand with the consequences discussed above.

Even in a controlled environment, the preferred response given above is certainly not the only option. You could offer to ship 25 early and the rest later, you could promise all 100 on the 2nd of April. The 2nd? Yes. Even though the current schedule shows only 80 available through April 2nd, this date is beyond CMLT so the plan can be increased with no impact on ongoing activity. The other consideration not shown on this display is production capacity. As discussed in Chapter 3, changes to the master schedule should always be verified through Rough-Cut Capacity Planning to be sure that the plan is within demonstrated capabilities.

Orders Within CMLT

Items planned for the "in-between" of this section title, that is, those shipped within CMLT, normally don't have to be manually consumed from the forecast. As long as the forecasted item is the same as the ordered item, the system should, in effect, consume the forecast automatically. This assumes that the system is driven by "greater demand" or "blended demand" which means that either the forecast or the backlog of orders is used in planning, whichever is greater. Out near CMLT and beyond, the forecast is usually all there is. As customer orders are received, the forecast is still the target, since not all customer orders have come in yet. Getting closer to the ship date, if things are going well, the backlog should approach the forecast quantity.

If backlog exceeds forecast within CMLT (the forecast was too low), you must decide whether to change the master schedule to accommodate the orders. Since a change to the schedule will be disruptive or might not be possible, this is usually an undesirable situation. One way to help avoid this problem is through the use of a safety requirement (safety stock).

Safety stock is a buffer against variation in demand. An item may have an absolutely steady *average* demand of, say, 100 units per month, but the actual demand in any given month can be expected to vary around that average. In those months when the variation is on the low side, there will be excess inventory which is applied to the needs of future periods. In the months when demand is higher than the forecast, there is a risk of running out of product. If the month-to-month variation is small, a small amount of safety stock will suffice. The wider the variation, the more extra inventory is required to provide the same service level (probability of being able to satisfy the demand). The same logic applies to an uneven forecast. Variation (actual demand compared to the forecast) is the issue, not the specific forecast model.

In an MRP system, a safety buffer need not be in-stock inventory. The safety quantity can be included in the plan one lead time in the future as a buffer against the amount of variation expected during the lead time. The planning system will consider safety demand during the planning cycle and, if needed, the extra quantity will be either in stock or in process depending on the lead time relative to the planning cycle. If the extra quantity is not needed because actual demand did not exceed the forecast, the activity bringing in the extra quantity could be deferred or the quantity could be "netted" against future projected demands. Placing the safety requirement one lead time in the future provides the buffer without risk of expediting simply to satisfy a safety requirement (safety requirement is never within lead time). In addition, the cost for this protection will be less than the cost of having the safety quantity physically in stock. The investment to cover safety requirements will be in work-in-process, on order from vendors, or sometimes some of it might actually be on the shelf, but shouldn't be for long.

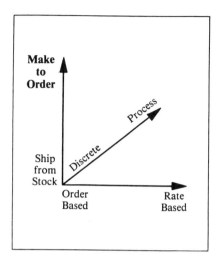

6. Make-to-Order and Government Contracting

A subset of the make-to-order community includes those companies that are true job shops in the narrowest sense, that is, those that provide manufacturing services and other short-turnaround, nonstandard activities usually selling to other manufacturers. They have no R&D of their own, they provide a service such as machining or assembly (although this discussion is certainly not limited to those two types of service) and often they are required to respond to customer requests immediately.

Another special case is the government contractor. In addition to the usual information management and operational control demands of manufacturing, this kind of company must comply with extensive reporting requirements mandated by the customer. Many times these requirements, at least in part, are also passed down to second- and third-tier subcontractors. The second part of this chapter addresses government contracting requirements.

Make-to-Order

Perhaps the two most telling characteristics of the true make-to-order job shop are the lack of data on which to plan, and the typically short turnaround time to complete the work that is often required. Both of these requirements pose a challenge to traditional MRP II operations.

MRP II depends on product definitions to perform its planning and management functions. The basic definitions include item characteristics, bills-of-materials, and process definitions (routings) in conjunction with facility characteristics. With the exception of the facilities, these data are usually not available to the job shop until the order is received or, at the earliest, when the prospect asks for a quote.

Planning, therefore, is problematic and the company is forced to use the kinds of assumptions outlined in the previous chapter such as planning bills and generic definitions. On completion of whatever planning can be done, the company must wait for the order to arrive and respond almost immediately on receipt. If the materials are available or are supplied by the customer, the work can start if there is capacity available. All planning is guesswork with a healthy dose of experience and intuition. The expeditious release of the order to the shop and efficient completion of the tasks must follow, and they are also hampered, from a system support perspective, by the lack of data.

Item Identification

Most order entry systems need an established item number before they can accept and record a customer order (which can then be passed to the planning and execution systems). While a "noninventory" item can be used in many systems, such an item is, by definition, undefined to the system and cannot be used in the other application areas. Usually, several departments are involved in establishing a new item (with all its characteristics such as classification codes, cost data, system control codes, etc.), and it often takes several days to gather all of the data and get them keyed into the system.

Several days of delay is intolerable in this environment, so another approach must be developed. Some companies establish skeleton records for new items which the order entry (or sales) people can call up and use as needed. Obviously, a skeleton definition is marginally useful as such and must be "filled out" as soon as the real item characteristics are known, and before the rest of the system can make full use of the item definition. In any case, at least a valid item number is immediately available for order entry to use, but procedures must be established and carried out to add the additional required information to make the item definition truly useful.

Some systems provide convenient capability to interrupt one process and switch to another with a few keystrokes, switching back just as easily to the original task. This allows order entry to establish a number but does no more than the skeleton approach as far as coordinating with the other departments and filling in all required information. Most companies would also be very uncomfortable with allowing order entry to establish new item numbers, which is a task normally reserved for engineering.

In some systems, item characteristics can be "copied" from an existing item to a new one. This feature can be used effectively if the person establishing the new item knows and understands the defining characteristics and can identify the closest match from which to copy. The basic definition is copied and the appropriate changes are made for the new item (same-as-except method). Group technology can help identify common characteristics. In fact, some companies establish generic items, sometimes using group technology codes as the item number, and use these generics as the basis for the same-as-except approach.

An even better situation exists if the customer has asked for a quote on the job. In preparing the quote, you would have access to the product characteristics and you could use your MRP system's database and standard costing facilities to help in the quote preparation. Once the data have been entered for the quote generation, they can be retained on the system and used when the order is actually received.

Some companies object to this suggestion because of fear that the database will become cluttered with definitions that will never be used. I believe that the use of the system's standard cost calculation subsystem is the best facility for preparing estimates (quotes) because it will be the most accurate (it uses the same definitions and process as your "real" costing system), it will be the most detailed (also because of the other definitions such as facility rates, etc.), and because it allows you to utilize same-as-except to simplify data entry.

Discipline is required to identify and purge definitions that are no longer useful.

Bills-of-Material and Routings

Once again, the lack of definitions hampers the quick availability of system support for a newly received custom order. Material cannot be allocated or even availability checked until the material requirements are defined through the bill-of-material. The same-as-except function is also the best solution available here. Predefined material lists, tied to similar items or generics, can be copied and modified as needed to speed up the entry of new item definitions. The same is true for the routing.

Fortunately, many job shops concentrate on a limited range of processes and products so that common materials can often be stocked for a relatively modest investment (sheet metal, rod, bar stock, etc.). In other cases, the customer supplies the material.

At least one MRP II system marketed to the "job shop" industry boasts of the ability to identify "noninventory" items at lower levels on the bill-of-material. Since these items are not identified within the system, a cost must be provided on the bill definition. These item requirements are passed directly to purchasing as "purchase-for-job" materials and are not recognized in the inventory system at all. While this approach may be convenient, there is always a price to pay for convenience.

The less definition and information in the system, the less management and control will be available. For example, if a purchase-for-job item is not used on the job, is received earlier than needed, or is purchased in a larger quantity than needed, there is no way to identify and store it except in the job-in-progress records.

Varying Bills

The last time I spoke before a group on the topic of "job shop" considerations, there was a company represented in the audience that made wire of various gauges. The process involves rod of various materials which is drawn down to the specified diameter. There are only a limited number of diameters and materials of rod that this company buys for stock, but the bill-of-material challenges only start there.

Quarter-inch rod (250 thousandths diameter) can be drawn down to many different diameters by a succession of steps, each yielding an intermediate diameter wire which could be a product in itself as well as a raw material for another, smaller size (see Figure 6-1). This company might take an order for a length of size C wire, which it would make from size A rod in two steps (0.250 to 0.100 in step 1, 0.100 to 0.085 in step 2). They could also draw down size B wire, if there was some in stock, in a single step. Although they do not normally stock size B, they might have some left over from a previous job, there might have been a customer return or an out-of-spec production lot that is suitable for reuse, or there might be an order in-house for size B of the same material and they will just overproduce on purpose to supply this new order.

I'm not aware of any packaged solutions designed specifically for this particular situation, but there are several procedural techniques that can be applied. First, the bill-of-material can be set up with each intermediate size as a phantom.* Most systems will check the availability of the phantom item in stock, and if none (or not enough) is available, it will "blow-through" to its components, in effect acting as if the phantom item was not there.

Playing with phantoms can be tricky, depending on the design and capabilities of your particular system, and might have implications for

---

*A phantom is a unique item type that designates an item that typically does not really exist. Originally the phantom item was created as a convenience for structuring a group of parts to a number of assemblies without redefining all members of the group each time. Most systems today allow considerable flexibility in what you can do with a phantom item. Often they can be built, stocked, issued, sold, etc., just like any "real" manufactured item.

MAKE-TO-ORDER AND GOVERNMENT CONTRACTING

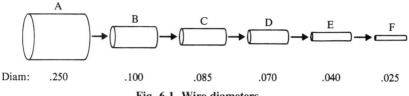

Fig. 6-1. Wire diameters.

the standard costing functions and the definition of routings. Check your system documentation.

Another possible solution is to establish a unique identification system that allows you to group several sizes under one part number. This works well with "off-fall" or the remainders of standard length rod, sheet, or bar stock.

Let's say that you use bar stock of a given dimension that ordinarily comes in 10-foot lengths. The standard unit of measure is linear inches. When less than a bar is left over after a production run, it goes into stock as a quantity equal to the number of inches remaining, but a short bar might not be suitable for all requirements. The next job might require 24 feet to make 4 each of the product for a total requirement (each product requires 6 feet). Two full bars (10 feet each) plus a 4-foot bar will not do.

If your system supports lot numbering, you can establish several lots for this product such as lot 2 (less than 2-foot lengths), lot 2–4 (lengths from 2 to 4 feet), etc. Although the allocation and the planning systems will not recognize the lengths, at least the picking process and visual records review will be able to identify what you really have. A similar identification system can be set up for the various wire diameters, but this is probably not as good a "workaround" as the phantom item suggestion.

Finally, the logic of your system can be customized to accommodate your unique needs. Be warned, however, that custom modification can be an expensive, frustrating process and the end result will probably be so unique that it will not be able to adapt very well as your business grows or changes. Refer to Chapter 10 for software modification/adaptation considerations.

Planning and Production Control

In most cases, the process is the important consideration (over material) since it is the service that is really being sold; so scheduling of plant facilities is of utmost concern. Scheduling requires a process definition (routing), and this information must be defined, as accurately as possible, before the system can be of much use in this area.

Once the requirements and the customer's requested schedules are known, the standard shop scheduling/priority planning facilities can develop a schedule and load profile (capacity requirements plan). Alternatively, a finite loading system can plan the resource availability and present any unresolved conflicts and/or material shortage considerations for resolution.

In addition to the finite versus infinite loading issue (discussed in Chapter 3), custom manufacturers are particularly interested in keeping all job information readily identifiable to the job and in coordinating schedules for all parts of a job.

MRP, in its most basic assumptions, tends to mix requirements together. Acquisition is planned according to need date without regard to the product or customer ultimately served. In the custom business, there is often a need or desire to be able to identify the ultimate use of every part, no matter how many levels down in the bill we may be looking. This ability to identify the customer or end product is called "full pegging," as opposed to the normal MRP pegging process which only identifies the next level parent items that created the demand. With normal pegging capability, the user must chase requirements level by level, matching dates and following the MRP logic in reverse to trace the top-level source of demand. Some systems aimed at the job shop market provide full pegging capability wherein each requirement in the planning system carries an end item or customer order identification. This requirement is similar to a "contract MRP" requirement you'll see later in this chapter.

One other aspect of MRP for custom manufacturers is an interest in scheduling forward rather than in the traditional backward MRP direction. Generic MRP starts with the master schedule (end item) due date and works backward until all supporting activities are planned. In this way, you get calculated start dates and completion dates for all purchase and manufacturing activities which, if executed on time, will result in an on-time completion of the product. The "custom" argument is that there is often insufficient time to do everything under "normal" conditions, in which case the MRP generation will tell you that you should have done a lot of things "yesterday." In addition, there is often a need to identify "when it can be complete," assuming we start today (or some other designated start date).

Forward MRP is actually pretty easy to implement. The basic methodology is to let MRP do its normal back scheduling, then call in a subroutine to identify the "offset"—the number of days between the entered start date and the first activity start date—and add that number of days to all calculated start and due dates. This shifts the entire schedule forward from the designated start date. Some packages offer an option to forward or backward plan; some even allow a choice by item or by job.

MRP assumes a known lead time for each acquisition (purchase or manufacturing) activity. While some systems allow a quantity-based (variable) lead-time component to be specified, it is still a "standard" time that does not reflect current load or conflicts. Using a variable lead time based on current load is an implementation of the finite loading philosophy (discussed in Chapter 3).

Another concern is to limit the movement of the "job" back and forth to inventory. A custom job is often considered to "grow" from materials through to the end product, whereas a multilevel bill-of-material represents a stage of production for each level. Normally, as each level is completed, the part or assembly is (at least logically if not physically) moved to stock and issued as a component to the next higher level. Some job shop systems will eliminate or automatically generate these receipt/issue transactions as activities are reported to higher-level stages of production.

Finally, scheduling of various components for a single job should be coordinated. If one component is behind schedule, there is no reason to rush other components that cannot be used until the late one is received. The tracking and coordination of "mating parts" is a special feature of some "job shop" variants of MRP. Without this capability, a late (critical) part must be traced (pegged) to its product, and the product's master schedule must be changed to agree with the part's availability. The next MRP generation would then reschedule all other parts and activities for that product.

Government Contracting

Manufacturing for the U.S. (or any other) government is a lot like manufacturing for any other big customer, with a few notable differences: the government is more demanding than most customers in terms of the specifications imposed on the contractor, the tracking and reporting requirements are generally more stringent, and government work is highly visible—there are literally millions of people with a vested interest in your performance from high government officials on down through ordinary taxpayers who are interested in how their money is being spent.

Scandals, whether justified or not, get big headlines when tax money is involved. We have all heard the stories about the $600 hammers and $1000 toilet seats. In truth, the requirements of the procurement process tend to inflate costs because of the tight specifications, the burdensome certification and reporting processes, and the typical low-volume nature of the production runs.

Nonetheless, many companies have survived and even prospered on government work, and it can be emotionally as well as financially rewarding to be involved in such important work. The key to success is to understand the customer's expectations and to be sure that you comply with all applicable requirements.

Fortunately, perhaps, the U.S. government is very specific about what is expected. The Federal Acquisition Regulations (FAR) run to many thousands of pages, and supplemental regulations and the details of individual contracts leave little to the imagination.

The "Ten Key Elements"

In 1987, in the wake of a revelation by the government that major deficiencies had been identified in MRP systems which could make contractors using MRP noncompliant with contract accounting requirements, the U.S. Department of Defense (DoD) issued a list of "Ten Key Elements" which they (DoD) considered essential for a system to be acceptable in a government contracting environment.

In an attachment to the "Key Elements," the government placed the burden of proof firmly in the hands of the contractor. If systems cannot be demonstrated to be in compliance, the government will withhold a portion of payments due "proportionate to the adverse...impact" and will restore full payment only on completion of an approved action plan to bring about full compliance with these key elements.

Several years after the "Ten Key Elements" were published, the government reemphasized its commitment to the principles contained in the "key elements" by making them a part of the Federal Acquisition Regulations (FAR) and thus even more binding upon government contractors. In any case, compliance with the wishes of the customer has always been and continues to be an important facet of the vendor–client relationship. Whether law, suggestion, or unwritten ethical standard, the customer expects certain things, and it is incumbent on the vendor to deliver. It is comforting, in a way, to have such a set of guidelines to specify what the customer's expectations really are.

In general, the elements address some basic system capabilities such as audit trails and consistent application of costs. Many of the elements also require, either explicitly or by inference, good "outside" disciplines, controls, and procedures which must be developed and maintained to provide the controls needed for accountability and verification.

The following discussion of the ten elements is my opinion based on my knowledge of MRP II and enhancement packages available and my experience as a government employee, a contractor, and a consultant to a number of prime and secondary suppliers to various government agencies including the Department of Defense (DoD). This opinion has not been verified by any government agency.

The Elements

The following is the list of ten key elements and comments to help interpret what they mean to an MRP II user company.

1. Have an adequate system description including policies, procedures, and operating instructions compliant with FAR/CAS criteria as interpreted by this guidance for all elements of affected cost.

Comments: While most packaged MRP II systems are quite well documented as to function and design, the requirements of element 1 really focus on the policies and procedures that must be developed by the user company to document how the system's capabilities are applied in their environment and what additional measures have been put into place to achieve control.

Any company, DoD contractor or otherwise, must develop and execute an implementation plan which adapts the capabilities of the MRP II system to the specific needs of the company. Inherent in this implementation is the assumption that the system is well designed, according to "standard" industry practice, and that not all of the functions of the system will be an exact fit to the current way of doing business. As different needs or requirements are detailed, the implementation team must decide how to address any differences between current practice and the industry standard as executed in the packaged software product. Whether the solution is a change in procedures, the writing of supplemental programs or additional reports, or the adaptation of the "standard" function through program changes, the actual approach must be fully documented and explained to provide a basis for consistent use and long-term stability in the application of the system's functions.

This "user" documentation should be sufficiently detailed to include the reasons for any changes or interpretations and, for DoD contractors, address the requirements of element 1. The government requirements are strict, and go beyond what most nongovernment users would consider adequate. The need for user documentation and recordkeeping is not unique to DoD and is ideally an interpretation or expansion of the complete "package" documentation provided by the software vendor. MRP II system documentation gives the DoD user a starting point for customized documentation to which he or she must add full documentation of how these functions will be used and what outside processes and procedures will be applied to complete the control system.

2. Assure that costs of purchased and fabricated material charged or allocated to a contract are based on valid time-phased requirements as impacted by minimum/economic order quantity restrictions. A 98% bill-of-material accuracy and a 95% master production schedule accuracy are desirable as a goal in order to assure that requirements are both valid and appropriately time phased. If systems have accuracy levels below those

above, the contractor must demonstrate that i) there is no material harm to the government due to lower accuracy levels, and/or ii) the cost to meet the accuracy goals is excessive in relation to the impact on the government.

Comments: Mention of "valid time-phased requirements" should reassure those who interpreted the 1987 headlines accusing "MRP systems" of contributing to overcharging that the government does, indeed, recognize the potential value of the MRP approach and has not discouraged or banned such systems. In fact, the first sentence of this element endorses time-phased requirements and economic order quantities (a succinct summation of MRP's process).

Part of the problem that caused all the furor in 1987 was that companies reimbursed for expenditures on a "progress payment" contract could accelerate the acquisition of materials to improve cash flow at government expense (the more material brought in, the higher the monthly bills to the customer). The phrase "valid time-phased requirements" is meant to address this issue and ensure that all material acquired is really needed in the near-term future and not merely brought in to enhance the billing amount. This element goes on to specify some parameters for the planning of these time-phased requirements.

The second sentence cites 98% bill-of-material accuracy as a *goal*, not a requirement. Any MRP implementor will concur that 98% is a typical target level for successful implementation of MRP. A company that is really serious about MRP will not stop there. Continuous improvement and an unwillingness to be satisfied with anything less than 100% accuracy are the marks of a successful project. In any case, the government states here that 98% bill accuracy is the expected level for determining valid requirements. If accuracy is less than 98%, it is presumed that there will be "harm" to the government unless the contractor can prove otherwise or prove that the cost of higher accuracy outweighs the "harm" caused.

This goal has nothing much to do with the system and everything to do with the attitudes, approach, and dedication of the implementation team. Some system functions can, however, contribute to the efforts. Bill-of-material accuracies this high can only be achieved if inventory reporting is done very completely and accurately, and if the company aggressively pursues accuracy by using actual usage to validate the "standard" bills.

There is a classic story, probably apocryphal, that tells of a group of executives discussing how a particular assembly was put together. When the discussion reached an impasse, the CEO, with the rest of the group in tow, headed down to the materials department to get the parts so that they could see for themselves.

MAKE-TO-ORDER AND GOVERNMENT CONTRACTING 89

When they arrived at the parts window, the CEO handed the bill to the attendant and told him to issue the parts. The attendant looked at the document, scratched his head, and said, "Now, do you want the parts on this list or the parts it takes to make this assembly?"

The point is that the people in the shop *know* what it takes. If they can be trained and properly motivated to report *exactly* what happens—what parts are used, in this case—and there is a feedback mechanism to compare actual usage to the bill-of-material, this offers your best opportunity to validate and maintain bill accuracy.

Today's CIM systems (see Chapter 9) address the conflict between the "engineering bill" and the "production bill." A basic ground rule for implementation of MRP II is that the bill-of-material must reflect material relationships as experienced in production. In order to properly plan material, the system's definitions must be as accurate as possible—not theoretical definitions but as used every day because this is how they will be applied. The measurement, therefore, is between the system definition and actual usage. Once again, timely and accurate reporting of inventory movement is the basis for measuring bill accuracy.

The issue of master schedule accuracy is more difficult to address. I'm not sure how one would measure this. Percent of shipments on time? Completeness of the master schedule as compared to contract requirements? Schedule within demonstrable capabilities?

I believe the intention here is, as stated in the balance of the sentence, to assure that requirements are valid and appropriately time phased. Whether this criterion can be audited or whether a sensible measurement can even be devised against which to audit for compliance are open questions at this time. Clearly, a valid master schedule is required for effective development of an MRP time-phased requirements statement. In line with this objective, the MRP II system should contain a state-of-the-art master production schedule planning facility which provides the necessary tools for developing and monitoring the master schedule. This master scheduling capability should tie directly to customer orders, to system-derived (or human-adjusted) forecasts, to a combination of these two sources of demand, to production-family aggregated demands (production planning), or under manual control. For validity checking, the system should include three levels of resource (capacity) checking: Resource Requirements Planning (item-group or family level), Rough-Cut Capacity Planning at the master schedule level, and a Capacity Requirements Planning module to analyze detailed work-center loading.

3. Provide a mechanism to identify, report, and resolve system control weaknesses and manual overrides. Systems should identify operational exceptions such as excess/residual inventory as soon as known.

Comments: It is interesting that the government expects the contractor to identify its own weaknesses, report them, and resolve any problems. As stated earlier, the burden of proof is firmly on the contractor. If an auditor should discover a control weakness that the contractor should have known about but didn't report, the contractor could conceivably be in violation of this element.

Manual overrides must also be reported. This could be a big problem for many MRP II users since overrides and file maintenance to correct errors are routine in many situations. To comply with this requirement, the contractor is expected to fully document and report any deviation from standard, documented procedures. The implication is that overrides should be rare exceptions and must be fully justifiable.

All MRP systems are oriented toward identifying and finding a use for excess materials. One of the basic calculations in the MRP process is "netting" requirements against available inventory to determine potential shortages and suggest acquisition activity to avoid these shortages. When "excess" inventory becomes available due to process improvements, schedule changes, etc., it is "used" to satisfy future demands, and subsequent acquisition activity will be replanned or postponed accordingly.

In compliance with element two, excess inventory should be made available to satisfy other requirements in accordance with normal MRP practice. If inventory was reported as acquired for a particular contract such as in a progress-billed or cost-plus contract, the system must allow transfer between contracts and transfer of the value associated with the items. Of course, all such transfers and adjustments must be properly tracked and reported in accordance with the requirements of the contract. The movement of material to and from contracts is addressed by several of the other elements listed below.

4. Provide audit trails and maintain records necessary to evaluate system logic and to verify through transaction testing that the system is operating as desired. Both manual records and those in machine-readable form will be maintained for the prescribed record retention periods.

Comments: As mentioned before, most MRP II systems are pretty well documented and, if installed and used at a number of other companies over time, have probably been audited and verified by independent authorities. Packaged applications, therefore, provide a good starting point for verification of system logic. Most packaged systems also provide good audit trails of transactions, file maintenance, etc., upon which you can build your case for compliance with this element.

You must include your own procedures in the documentation, however, to verify that the way *you* apply these functions is in compliance with the requirements of *your* contract. You are obligated to test and verify, with transactions, that the system is operating according to your expectations, and you must maintain auditable records.

MAKE-TO-ORDER AND GOVERNMENT CONTRACTING 91

A more important consideration might be how well the users understand and apply the system functions, and what manual procedures have been developed to ensure data accuracy and compliance with government requirements. As discussed under element one, complete "user" documentation is required for auditable use of the system. Auditors will be very interested in how day-to-day activities support the system-provided functions, and whether security functions are regularly tested and monitored.

5. Establish and maintain adequate levels of record accuracy, and include reconciliation of recorded inventory quantities to physical inventory by part number on a periodic basis. A 95% accuracy level is desirable. If systems have an accuracy level below 95%, the contractor must demonstrate that i) there is no material harm to the government due to lower accuracy level, and/or ii) the cost to meet the accuracy goal is excessive in relation to the impact on the government.

Comments: Ninety-five percent inventory accuracy is a reasonable goal and is generally accepted as the required level for successful implementation of MRP. It is generally accepted that the only way to achieve this level of accuracy is through procedural discipline and the aggressive application of a cycle counting program.

Note the wording in the first sentence. The element specifies accuracy reconciled by part number. The total dollar value of a physical count as compared to the balance records is not good enough. Counts can be wrong in both the positive and the negative direction, and these errors tend to balance each other out. What the government is looking for is what most serious MRP II users recognize as the "correct" measure of accuracy—the count accuracy by item.

If you count 100 items in inventory, how many are correct? It doesn't matter if the count is off by a little or a lot—the count is either right or wrong. This stringent measurement of accuracy is necessary to ensure availability of parts and usability of MRP.

Once again, the contractor must prove that he or she has maintained this level of accuracy, or either prove that the inaccuracy does not harm the government or that the cost of compliance would cause greater "harm" to the government.

6. Provide detailed descriptions of circumstances which result in manual or system-generated transfers of parts.

Comments: Transfer of parts to and from contracts and between contracts is a big concern to the government since there is ample opportunity in parts transfers to mischarge, overcharge, or lose accountability. Several of the following elements address specific circumstances of parts movement and how they should be handled. The purpose of this element is to require full documentation of why transfers occur, whether system instigated or manually prompted, so that the auditors can verify that the transfer was justified and in compliance with all of the rules contained in the specific direction of the following elements.

7. Maintain a consistent, equitable, and unbiased logic for costing of material transactions. The contractor will maintain and disclose a written policy describing the transfer methodologies. The costing methodology may be standard or actual cost, or any of the CAS 411.50(b) inventory costing methods. Consistency must be maintained across all contract and customer types, and from accounting period to accounting period for initial charging and transfer charging.

(a) The system should transfer parts and associated cost within the same billing period.

(b) In the few circumstances where it may not be appropriate to transfer parts and associated cost within the same billing period, use of a "loan/payback" technique must be approved by the ACO. When the technique is used, there must be controls to ensure that i) parts are paid back expeditiously, ii) procedures and controls are in place to correct any overbilling that might occur, iii) at a minimum, the borrowing contract and the date the part was borrowed are identified monthly, and iv) the cost of the replacement part is charged to the borrowing contract.

Comments: The transaction-reporting disciplines required by a computerized system typically are much more restrictive than in manual systems which are not based on well-defined logic. Therefore, it is reasonable to expect that transactions can be more accurate in an automated system. Further, most computerized systems will apply costing logic at the time the transaction is processed so cost accounting should be more timely than in a manual system. Computer logic will perform costing functions consistently as long as the users do not override or somehow circumvent the controls built into the functions. The previous elements addressed the need to document any overrides or manual intervention. Material transfers will be documented (see the following elements for specific requirements).

This element states that a number of different costing methodologies can be used. The specific method required might be specified in the contract, or the contractor may be allowed to choose a method and provide the audit agency with documentation of the chosen methodology and how it will be consistently applied. The one requirement here is that all customer and contract types must be treated consistently, and the method may not be changed (period-to-period consistency).

Point (a) specifies that transfer of parts and costs must occur within the same period. This is to avoid a situation in which two contracts might be billed for the same item. If the item is transferred, the cost must also be transferred within the same billing cycle.

Point (b) goes on to say that if there is not a complete transfer of part and cost within a billing cycle, the methods used to manage the situation must have been preapproved by the auditing agency and must meet the listed criteria. Perhaps the most interesting aspect of borrow/payback is the requirement to charge the replacement cost to the borrowing contract.

In the presence of "valid, time-phased" requirements, parts should be acquired only when needed. A situation can occur, however, where parts are in stock and charged to, let us say, contract A, but an unexpected requirement appears for contract B which has a higher priority (earlier requirement date). The nature of MRP is to satisfy the earliest requirements first, so A's material is made available to contract B, the borrower.

If replacement is made to contract A within the billing cycle, no accounting is necessary, just documentation of the loan and its repayment. If not repaid by the end of the billing cycle, however, contract A must be credited for the value of parts it no longer controls and the cost must be transferred to contract B.

Let us say that the original parts cost $100. The replacements, however, cost $125. When the replacement parts are brought in for contract A, the contract is recharged the original $100 and the charge to contract B is changed to the $125 replacement cost. In this way, contract A is not penalized for loaning its parts to another contract.

8. Where allocations from common inventory accounts are used, have controls in addition to the requirements of elements 2 and 7 above to ensure that:

(a) reallocations and any credit due are processed no less frequently than the routine billing cycle;

(b) inventories retained for requirements which are not under contract are not allocated to contracts; and

(c) algorithms are maintained based on valid and current data.

Comments: It is not unusual for a company to maintain an inventory of common parts which can be used on various contracts. The costs for this inventory are determined using the approved accounting methods and are applied to the contracts based on valid, time-phased requirements. This element seeks to ensure that items not purchased for a specific contract are handled in much the same way as those that are.

Items issued from "stock" inventory can be charged to a contract when appropriate. If some are returned to stock or transferred to another contract, the same accounting requirements apply as would apply for other items transferred. Extra quantities purchased should not be charged to a contract unless specifically authorized and, if they are authorized and charged, they cannot be used on another contract unless the appropriate transfer of costs occurs.

9. Have adequate controls to ensure that physically commingled in-

ventories (which may include materials charged or allocated to fixed price, cost type, and commercial contracts) do not compromise requirements of any of the above key elements.

Comments: This element endorses the concept of commingled physical inventories with segregated costs.

The government doesn't really care (in most cases) if the actual item purchased or made for a contract is specifically retained and identified with the contract or is used on another contract, as long as the "part" is available when needed and charging is fair and equitable. You may replace an acquired part with an identical substitute with no penalty and no tracking unless the situation falls under one of the preceding rules. Serial number controlled items or those subject to unique certification requirements may not fall under this provision. This is basic to the application of MRP in a contracting environment.

The principles of MRP revolve around the concept of availability: an item is made available to the earliest need and replenishment action is tied to needed quantities and dates. There is no identification, within generic MRP, of planned or completed acquisitions to their ultimate usage, except within the mathematics of the planning calculations. As requirements change, the calculations change and parts can be reallocated as needed. Therefore, the parts originally planned to meet the needs of contract X might well end up in contract Y. It is only when costs are assigned to a contract that accountability becomes important.

Parts acquired to meet the needs of multiple requirements can be brought in as common inventory (element 8) and allocated at the time they are physically assigned to a contract. Alternatively, the costs can be allocated at the time of acquisition (subject to the terms of the contract), but cost tracking must follow physical movement of the parts within the requirements of elements 6 and 7 as in the case of loan/payback.

There are systems on the market that accommodate "contract MRP" or "MRP by contract." This might be a contradiction in terms since MRP is designed to ignore the ultimate use and to focus strictly on availability. In any case, contract MRP performs the same function as regular MRP but limits the calculations to the requirements and inventories of a single contract at a time. You might be able to accomplish the same results by establishing in your system a separate "warehouse" for each contract and running separate MRP generations by warehouse, if your system supports that capability.

As always, the contractor must provide good audit trails and be able to verify compliance with disciplines and procedures instituted to support the required accountability.

10. Be subject to periodic internal audits to ensure compliance with established policies and procedures.

Comments: The final element reemphasizes a theme that is consistent throughout the key elements—the need to establish clear, written procedures and to be able to verify compliance with these procedures on a routine basis. Systems must be designed to handle data and provide audit trails as outlined; but without good outside controls and an aggressive internal commitment to verification, the best system controls in the world can't be effective.

Once again, the burden of proof is placed squarely on the shoulders of the contractor. You must, through your own efforts, verify that your procedures are sound and that they are being upheld. The auditors are within their rights to ask to see records of verification that procedures are being followed as documented and approved.

General Note: Although government acquisition regulations cover literally thousands of pages of text and include overwhelming detail, the requirements of each individual contract will vary from every other. Compliance is in the eyes of the customer for each contract and is verified by the designated audit agency. The best advice I can offer is to understand what your specific contract requires and to meet with the auditors to find out specifically what they will be looking for. Most contracts require that the cost tracking systems be described and approved before significant charges are made against the contract. I encourage you to exploit this opening, at the earliest opportunity, so that the systems and procedures that you set up will in essence be "preapproved" to avoid difficulties at a later time.

Additional Considerations

In addition to these ten general elements, each contract will specify what data are required and how they are to be reported. Most contracts require that cost and schedule information be reported monthly, in a specified format, and be organized according to the contract structure.

Contracts are often subdivided into a hierarchical structure including major divisions (often called phases), subdivisions, sub-subdivisions, etc., the extent of which is determined by the size, complexity, and duration of the contract. This hierarchical arrangement of contract reporting elements is called the Work Breakdown Structure (WBS). The requirements of the contract are arranged according to the WBS, as are the shippable products, supporting data deliverables, testing and certification, engineering tasks, and all other activities which the customer specifies. Reporting of schedule status, as well as costs, must be subdivided according to the WBS.

Unfortunately, a generic MRP II system will gather costs primarily by work order, and individual work orders must be associated with the

WBS elements to which they apply in order to accumulate progress information in the required format. There are several MRP II systems on the market that have either been developed with this environment in mind or have been extended to address these reporting requirements. In most cases, the contract data are an accumulation of information that already exists in the system into increasingly higher-level summaries, so the adaptation of the package is primarily one of additional reporting facilities rather than changes to the basic logic of the system.

Other supplemental requirements include, of course, a facility for storing the WBS structure, maintenance of budgets or estimates, comparative reporting capabilities (actual versus authorized budget), and the ability to project the cost-to-complete based on current status.

Priority Work

In times of crisis or for critical needs at any time, the government reserves for itself the right to demand priority treatment at the expense of all other customer requirements. Commonly called a "DX" priority rating, an order carrying such a designation must take priority over all other commitments. Generic MRP does not recognize priorities other than date of need. By placing the DX order in the master schedule with the earliest date, its needs will automatically take precedence over all others. If the priority order causes an overload of resources (either capacity or material shortage), the planner and the functional managers must ensure that the priority job gets the resources first.

Sometimes there is not enough time to regenerate the plan. In an "hour-based" response situation, the user must rely on shop scheduling systems and availability checking functions rather than the planning system to execute the order.

Configuration Control

Configuration management should be a concern for all manufacturers, but tracking of configurations over an extended period of time is a common requirement in government work. The vendor is often required to maintain a history of specific bill-of-material used for each product unit or lot. This requirement is most easily satisfied with an archiving capability for production (assembly) activities tied to the serial number or lot number of the product.

In addition, revision level control is a collateral requirement, especially in an engineering-driven process. If the product is being designed or improved as it is being produced, control of the engineering release process is critical and must be linked with production to update bills and revision levels in-process as well as to ensure traceability as the design evolves.

While there are add-on engineering release control packages available for many packaged systems, the ultimate solution is tighter integration between the business and planning system and the engineering department functions. Specifically, CIM systems such as IBM's CIM Architecture bring product descriptions (item definitions, bills-of-material, routings) into a centralized facility where they are shared by all applications that need them, rather than being synchronized because of duplication. A CIM system should have a release control subsystem which allows product-in-development definitions to be managed until release when they are loaded, en mass, to the centralized "official" definition and synchronized with work in process. Further discussion of the CIM Architecture is included in Chapter 9.

Project Management

While this requirement is somewhat outside of the normal range of MRP II functions, there is usually a need to track nonmanufacturing activities such as engineering, testing, and documentation work. Many companies establish work centers and use work orders for these tasks as if they were "real" production. A typical MRP II shop-order tracking system works splendidly at these tasks, providing a mechanism for reporting and tracking time, labor charges, materials used, and progress toward a schedule.

In addition, most contractors use a packaged project management system, typically PC-based, to maintain milestone information, PERT and Critical Path (CPM) charts, and related schedule information to satisfy contract reporting requirements. The challenge is to integrate the project management format with MRP II production cost and status information to satisfy the overall reporting requirements with as little duplication of effort (and data) as possible.

There are (of course) extensive government specifications that cover the reporting requirements and formats. If the contract is big enough, it will fall under a specification known as C/SCSC, which stands for Cost/Schedule Control System Criteria, affectionately called "C-spec."

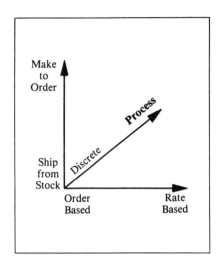

7. Process Manufacturing

The simplest definition of process manufacturing that I have come across is: the making of products that pour. Being a little more specific, process industries work primarily with liquids and powders; mixing, cooking, and combining the materials into products often relying on chemistry and physics to enact the transformation from materials to products. By contrast, in discrete (the opposite of process) manufacturing, the change from raw materials and component parts (rather than ingredients) to products is normally as a result of the application of labor or machine activity, in other words, the value is added through outside action rather than internal chemical or physical processes.

Of course, there is room for overlap in these definitions. Many manufacturing processes incorporate elements of both discrete and process, and the distinction is more important in the context of this discussion of techniques and software functions than it is in reality, on the plant floor.

Some common characteristics of process industries include: relatively flat bills-of-materials (usually called formulations, formulas, or recipes); so-called "upside-down" bills (one or a few components yielding more than one product); the need to recognize fractional units of measure (transactions and balances carried to decimal places of the basic unit of measure which is likely to be a measure of weight or volume rather than "each"); concern with yield, potency, and grading; short production processes; large production quantities; and production capacity being more of a limit than material availability. Of course, these characteristics are not strictly limited to process industries, and not all process manufacturing entails all of these concerns. In general, however, these are some of the major concerns of process manufacturers.

Process manufacturing can be addressed by basic MRP functions, although there are often severe limitations, in addition to the differences in terminology, that can limit success or challenge the users' creativity in their application.

"Batch" Formulas

One common characteristic of process manufacturing is the need for more precision in formulations (the equivalent of bills-of-material) than is commonly provided in a basic MRP system. It is not unusual for process batch sizes to be large in relation to the specifications for one unit of the product and thus present a challenge in specifying the units of the components. One aspirin, for example, might contain 325 milligrams of active ingredient, but you would never make aspirins one or a few at a time. If a production batch size is 100,000, for example, the material pick list would specify 32,500,000 milligrams of this component. A scale that can handle 30+ kilos is probably not accurate to the nearest milligram, and material handlers would probably not be happy with all the unnecessary zeros.

If you could specify the ingredient in kilograms, it would be much more convenient for the material handlers. The formula would then state that each aspirin requires 0.000325 kg. Unfortunately, most systems do not have enough precision in the formula specification system to handle that many decimal places.

What many process companies have done is translate units of measure either off the system or through the use of phantom items. The off-system approach assumes that everyone will interpret the data on pick lists, formula reports, etc. The formula might say grams, for example, but everyone knows that it is really kilos (wink, wink). Don't laugh! I've seen this kind of thing in a number of companies. One company I know specified all formulas by weight—powders, liquids, everything. This was not only hard on the material people (converting all incoming units to grams, weighing liquids for issue to production), but the lab wasn't too happy about it either.

The usual solution to this problem in a "process-oriented" MRP package is to provide the ability to specify formulas in terms of a quantity other than one unit of the parent item. This facility is referred to as a "batch bill-of-material" capability. A batch quantity is specified for the parent item, then the quantity-per for the direct-use components is stated per batch of the parent rather than per unit. The product can be made in any desired lot quantity and the system will apportion the component quantities appropriately.

The standard batch quantity is assigned for each item as needed. Our aspirin example could be specified at a quantity of 100,000, for example, and the ingredient quantity-per would now be 325 kilograms. The pick list for a lot of 50,000 would include 167.5 kilograms of the ingredient, which is much more preferable to 167,500,000 milligrams or some other compromise solution. The standard batch quantity of the ingredient, in turn, might be 500 kilos, or any other convenient quantity without regard to the batch quantities of any other item.

Coproducts and Byproducts

The so-called upside-down bill-of-material is another challenge for a traditional MRP system. In most systems, as in discrete manufacturing situations, it is expected that the bill will describe the components required to make a single end item. There is no provision for a second or third product resulting from the same process.

Coproducts and byproducts are common in process environments, although not unknown in discrete situations (recoverable scrap such as the remains of copper coil after parts have been stamped out of it). In addition to planning other useful outputs from a process, hazardous wastes can be identified as byproducts so that their disposal can be planned.

A process-oriented package might allow more than one item to be designated as a product of a formula by using an item characteristic that designates it as a coproduct or byproduct. The standard "fix" in systems without this capability is to allow components with a negative value for the quantity-per in the bill definition. Since positive quantities are allocated to a production requirement, a negative quantity can be negative allocated, reflecting an expectation of a negative issue which is the same as a receipt.

This backward logic may sound a little crude, but it works. Using a discrete example, the bill-of-material in Figure 7-1 illustrates the situation for a stamping process wherein sheet copper is the primary raw material and copper disks are the product. After the disks are stamped out, however, there is considerable scrap copper left over. This obviously has value, should be recovered, can be recognized as a byproduct of the process, and should be reflected in the planning and costing systems. The negative quantity-per allows the planning system to anticipate receipt

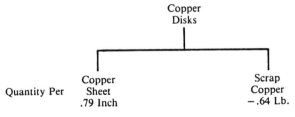

Fig. 7-1. Byproducts in traditional MRP.

of the "negative issue" quantities, and the costing systems will recognize the value of the byproduct as a negative cost element.

In addition to the rather confusing logic of this solution, there is the additional difficulty of not being able to master schedule the byproduct since it is defined as a component of the process, not the product. That's probably not a problem with the scrap copper, but it might be a significant drawback in the case of a coproduct, a more significant additional output of the production process.

Packaged solutions that are enhanced for the process industry include a code in the item definition and/or the bill-of-material that identifies byproducts and coproducts as distinct from products. The software then recognizes the special characteristics of these items and allows them to be identified correctly rather than using the back-door method outlined above.

Yield

Many packaged MRP systems today allow the addition of a yield factor in the description of the process (the routing). Yield is the loss of some of the (partially completed) product quantity during the process. Yield losses, while certainly not limited to process applications, are quite common when working with volatile liquids, with heating processes, or when mixing ingredients together and are a result of spillage, evaporation, mix clinging to utensils and containers, and so forth.

Since yield reflects a decreasing batch size as the process progresses, any ingredients added after the yield loss will be in a proportionally smaller quantity than components added before the loss. The manual adjustment to quantity-per can be done in the bill-of-material specification easily enough, but successive yields in more than one step can complicate the math, and dealing with cumulative percentages is sometimes cumbersome. If the user can specify the expected yield losses as percentages in the operation steps where they will occur, the software can perform the calculations to scale the component quantities accordingly, and also can capture actual yields for comparison to the expected (standard) yields as illustrated in Figure 7-2.

Many off-the-shelf MRP solutions now contain a yield factor in the routing definition. A program calculates adjusted quantity-per for components tied to where they enter the process, and reported actual yields are compared to expected yields.

As a secondary consideration, an adjusted quantity-per can be helpful in the case of an ingredient with a limited quantity available. Most systems with a yield capability also allow the user to specify an "in" quantity of a component, and the system will apportion the other mate-

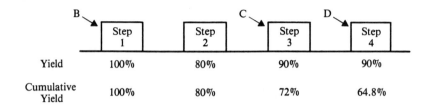

Fig. 7-2. Yield.

rials and calculate the appropriate batch size to "use up" the specified ingredient.

Potency

Process ingredients often exhibit variability that affects their use in the process. Especially with "natural" ingredients, the strength or potency of the material affects the amount needed. Traditional MRP bill-of-material processors have a problem with this because they have no way to identify and apply whatever conversion factors might be necessary to accommodate this variability. The acidity of tomatoes, for example, might vary based on the weather, time of the season (early harvest versus later crop), tomato variety, etc., and the amount of material needed for a particular recipe might vary with the acidity. In addition, the amount of sugar or other additives (compensating ingredients) needed in the process can be dependent upon the acidity of the vegetable.

Some packaged systems use a code in the formulation to identify the active ingredient and (optionally) the compensating ingredient so that the program can calculate the ratios for you. There is also sometimes a strength or potency code that allows the formula processor to adjust the needed quantity based on the reported strength of the component. This can be done by either including a factor (multiplier) used in conjunction with a percentage-type potency factor or through the classification of strength ranges into grades. A component might be specified as grade A,

B, or C, for example, and the bill would allow different quantity-per requirements for each grade:

Grade	Quantity-Per
Grade A 90% +	1.0
Grade B 60% – 90%	1.4
Grade C 40% – 60%	2.0

Grading

Grading considerations also apply to products as well as ingredients. When making a process product, you don't always know what you will end up with. You might set out to make Grade A Whosit Juice but actually produce Grade B instead; or you could end up with 70% Grade A and 30% Grade B. The existence of a coproduct is not so much of a problem in itself (see discussion earlier in this chapter), but anticipating and planning for the variation adds to the challenge.

Process-oriented solutions sometimes include provision for grading of coproducts, and planning systems are set up to recognize expected variability based on a forecast. This part of the process is really not much different from the use of a planning bill to anticipate customer requirements for a variety of products within a product group or family.

As far as workarounds for potency and grading, there are no standard methods that I have seen used extensively. Each company decides on how to address these concerns based on their specific situation. Don't interpret this statement as an admission that this is an unsolvable problem. There is enough variation in the requirements that general rules don't really apply. Suffice it to say that use of standard formula definition functions can be applied to potency and grading concerns, but these workarounds are usually limited to specific needs.

Back-Flushing

Back-flushing is a common methodology in process manufacturing used to automatically generate issue transactions for components when production is reported, using the quantity produced and the "standard" quantity required for the materials. Especially in high-volume flow production, it is normal practice to stage materials at the appropriate location near the line and "assume" usage based on the quantities of product produced. With the typical process liquid and powder ingredients, feed pipes, chutes, and conveyors are often used to transport materials to the point of use eliminating the opportunity to perform issue activities and transactions in the normal sense.

Back-flushing is useful under the above conditions to account for the expected amount of material used, but the technique is frequently overused and misapplied simply to reduce the number of transactions required to "feed the system." The danger is in the assumption of normal usage. With no actual reporting, there is no opportunity to report and identify variances other than by after-the-fact reconciliation of remaining quantities (we should have 200 pounds left but there are only 180, therefore we must have used 20 more than expected; now, which products use this ingredient and which one(s) had the variance?).

In a situation where ingredient input cannot be conveniently measured, the back-flushing assumptions are acceptable and useful. Where back-flushing is being used simply to avoid transaction entry, the inaccuracy and lack of feedback are a high price to pay for convenience.

As a convenient alternative to back-flushing, many systems today allow material issue transactions to be entered on a "complete except" basis. The user identifies the production order or schedule, and the system displays the required components, sometimes with availability information as well. The user can accept the complete list as is or can specify exceptions to the standard quantities for one or more components, then accept the rest. This is far preferable to back-flushing because the user has a convenient method for reporting exceptions that is lacking in a back-flush situation.

Since many process situations involve high-speed high-volume automated machinery, there is often a desire to back-flush labor and overhead as well as material. More discussion of this topic is included in the "flow" manufacturing section in Chapter 4.

Other Process Requirements and Features

Process manufacturing is often done on a large scale using continuous flow processes. Thus, many process companies are also interested in the MRP extensions addressed in Chapter 4 to support rate-based work planning and management. In addition, there are other needs in peripheral areas such as warehouse and inventory control, costing, material acquisition, and planning, and quality (process) control.

Batch/lot traceability is often required in food, chemical, pharmaceutical, paint and pigment-related industries, and other process environments. This need requires a capability in the inventory control system to attach qualifying information to inventory balances as well as the ability to track multiple lots of a product in multiple locations. The added information will identify the batch/lot or control number, when the product was received, its QC status (usable, awaiting inspection, rejected, expired), and when further inspection or expiration will occur.

Supporting functions to apply and maintain these characteristics are also needed, such as the ability to designate an item as not usable until QC approved, tracking and disposition support for shelf life, etc.

Another area in which process requirements might differ is in the general approach to cost accounting. Since there is often no "order" against which to log activities, costs are accumulated by department or work center by time period (shift, day, week, month), which is often called process costing or departmental costing. No attempt is made to attribute costs to specific run quantities of specific products, as in job costing, since this would significantly increase the reporting and tracking requirements beyond what is practical in a fast-moving situation.

Material acquisition considerations in process manufacturing tend to be more advanced than in discrete manufacturing primarily because the scheduling of incoming materials is often more critical. When the raw material comes in a 45,000 gallon rail car and the car must be emptied on the day it arrives, schedule and delivery dates are less forgiving than if the item can be placed in the warehouse until needed.

Process companies, therefore, tend to take full advantage of comprehensive purchase tracking capabilities including extensive management and reporting functions, vendor performance analysis capabilities, and facilities for handling blanket orders and varying "releases" of blanket orders. The purchasing system must, of course, fully integrate with the inventory receiving, lot control, and QC functions mentioned above.

In the planning area, production planning—the ability to group items together into "families" and plan in the aggregate—is particularly useful in many process situations. This applies to product variants such as colors, strengths, various package sizes of the same product, and a number of products produced on the same line. In conjunction with aggregate (production) planning, resource requirements planning and, at the master schedule level, rough-cut capacity planning are also popular functions for process companies. Master scheduling is discussed in Chapter 5.

Quality and process control considerations are customarily addressed as add-on functions outside of the basic MRP system. Conceptually, MRP doesn't really concern itself with tracking quality information. MRP's purpose is to use information about the status of things to plan operations within that set of parameters. In other words, if you tell the system to expect 10% unusable parts, it will plan to acquire 10% more so that you will not run short.

Most companies today are very concerned with tracking and managing quality. The process industry, in particular, is often subject to regulatory oversight as well as a concern for the quality of their products and the satisfaction of their customers. Parametric measurements (viscosity, pH, specific gravity, other test results) and statistical analysis are customarily handled by a separate module with little direct interface into the MRP system. Some MRP vendors offer quality modules, some are available from third-party vendors, and some companies create their own.

Shop Activity Planning and Control

Many process manufacturers are a combination of discontinuous production and continuous flow. Typical of the breed is the "mix and pack" situation, in which a batch of product is produced in a vat, tank, or other vessel (discontinuous), then piped to automated packaging equipment (continuous or flow production), thus requiring a mixture of the techniques outlined in the previous chapters. Usually, it is the packaging line that is the constraining resource, and all scheduling is oriented around the limitations of the line (line speed, units per hour) and whatever sequence requirements there might be. The batching portion of production is slaved to the needs of the packaging line just as feeder lines can by synchronized to the needs of assembly lines in discrete flow manufacturing. Packaging materials are often loosely controlled (stocked in relatively large quantities) so that they will not unduly influence the scheduling of the line.

Of course, there are exceptions to any of these generalities. I once worked with a winery that had very limited storage space, thus they scheduled delivery of bottles very carefully. The empty bottles that were delivered in the morning left that same day for off-site storage or delivery to customers full of wine. There was no other choice. If the bottles didn't come in on time, this company was forced to either wait, with the equipment sitting idle, or change the schedule to something else for which they had containers.

Other than these special situations, production control considerations are as addressed in the other chapters on "normal" production, continuous flow production, and the consideration of a job-shop (custom) manufacturing situation, since process manufacturing can fall into any one or more of these categories.

An Alternative Approach

Marcam Corporation (Needham, MA) has developed a software alternative to traditional MRP II specifically for the process industry. This product, called Prism, does not use conventional bills-of-material and routings to describe the products and processes, but rather contains a patented approach they call the "Process Model" which takes the place of

PROCESS MANUFACTURING

both. The following description of the Prism system is an example of an unconventional approach to production planning and control that does not follow the basic design and assumptions of traditional MRP.

Essentially, the process model allows the user to define needed resources—whether materials, machines, people, utilities, or anything else—as well as products, coproducts, waste, and any other output, on a single-level basis. Inputs and outputs are treated in the planning and control systems on an equal basis, allowing the planning for materials and facilities at the same time, and subject to the same rules and conditions. This effectively eliminates the arguments about material-first/resources-after, versus the opposite sequence, that plague the MRP II community (see Chapter 3).

The process model also solves the upside-down bill-of-material issue rather neatly while allowing convenient handling of other output such as waste water, recyclable scrap, and/or hazardous waste.

Marcam describes a sample situation using the diagram in Figure 7-3. I find it interesting that they draw their diagram from the bottom up, rather than the usual bill-of-material and routing diagrams that read from top to bottom. Maybe they are trying to tell us something?

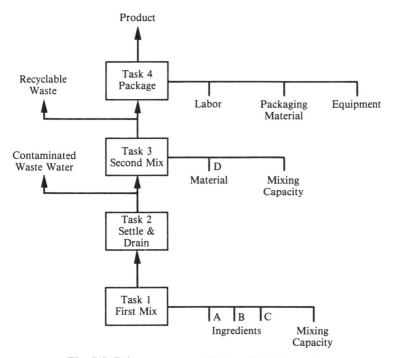

Fig. 7-3. Prism process model (copyright Marcam).

The Prism Process Model is designed around a basic definition of resource as anything with value, positive or negative. Therefore, anything that is added to a process or results from a process can be a resource, can be defined within Prism, and can be tracked and managed. Any resource can also have limitations whether a balance-type positive resource such as materials, or a nonbalance negative resource such as waste water. Your process may be limited by the amount of waste water you can handle in a day.

Once resources are defined, they are structured into the Process Model according to the tasks at which they are consumed or produced. In Figure 7-3, three materials plus some mixing capacity are added to task one. At task two, waste water (a byproduct) is produced. At task three, more material and mix capacity are added. Finally, packaging consumes some more resources and produces the product plus a byproduct.

Since all resources are handled at the same level in the planning system, the availability of mixer capacity, labor, handling capacity for byproducts, the raw materials, and the packaging can all be planned or can limit the production process.

Prism is extremely flexible. In the inventory subsystem, for example, an item can be "inbound," on-hand, at QC, rejected, or in transit. It may be available, consigned, or on hold. There is also a series of "yes or no" choices as to whether it can be available, ordered, costed, in WIP, lot controlled, or planned (now or later). Inventory transactions are also user-definable by class or subclass of item. Tasks can contain multiple steps which can be reportable or not (back-flushed). Reporting can be done at the step, task, or only at the end point. You can also report in aggregate and have the system apportion any variances to all tasks completed during the reporting period.

Prism is so flexible, in fact, that it has been criticized as difficult to implement. Each resource and activity must be fully defined and identified so that the system will know how to handle it, and the approach is so unique that there are no other similar situations to use as a guide outside of the Prism community (past MRP II experience is only marginally helpful to a new user).

This is an illustration of the normal tradeoffs that face software developers and users in any environment: the more flexibility that is built into the system, the more the users will have to think in order to use it correctly. Fully implemented users report that changes to the definitions to reflect changes in the process are easy to enter and use, allowing Prism to adapt to complex and changing manufacturing environments.

Since Prism was designed specifically for the process industry, it also contains just about every feature discussed in this chapter plus some others, such as an integral activity based costing capability and a cooperative processing extension using a graphical interface.

Summary

Process manufacturing *is* different from discrete manufacturing in many ways. Most of the major limitations of a packaged solution designed for discrete situations have been overcome by enhancements built into successive versions of most major packages. Specifically, the batch formulation capability, yield features, QC tracking functions (quarantine, shelf life, rejected material, batch/lot control), and repetitive production management features are readily available as standard features or add-on modules for many systems.

Some of the requirements that are not well addressed in most packages include potency considerations, coproducts/byproducts, and product-specific needs.

As with any packaged software decision, the most important thing is to understand your needs first, then compare the relative merits of various software offerings. More function and more flexibility necessitate more complexity and, usually, higher cost. You should identify how well a particular package fits your specific needs, understand any shortcomings, and have a plan for how the missing features or functions will be handled. Chapter 10 discusses adaptation and modification of packaged software.

8. Activity-Based Costing

Years ago, it would have been typical for cost-of-goods to be made up primarily of labor content with relatively smaller amounts of material and overhead. In preautomation days, 60–70% labor content, 10–20% material, and 10–20% overhead would have been normal. With the introduction of automation, the labor content of the typical manufactured product has decreased dramatically. Today, the typical breakdown is 60–80% material, less than 10% labor, and around 20% overhead.

Traditional cost accounting systems usually apply overhead as a percentage of labor or a labor-related factor—labor hours, labor dollars, or activity (machine) time or dollars. The old distributions of material, labor, and overhead would result in overhead rates of 10–40%—in other words, labor, which was 60% of the content would drive about 70–90% of the cost (labor's 60–70% plus the 10–20% overhead "burden").

With current cost distributions, driving overhead through labor or labor-related factors results in overhead rates of anywhere from 100 to 600 or 700%—a few percent of the content now drives up to 30% of the costs or more. (See Figure 8.1.) If small errors are made in the labor accounting, large errors result in the costing system. It is the stereotypical tail-wagging-the-dog situation.

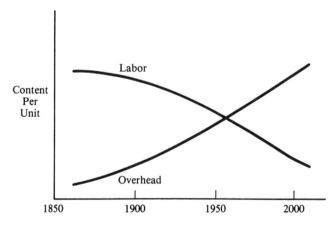

Fig. 8-1. Cost-of-goods content.

Because overhead is seen as an uncontrollable factor, there is usually no attempt to understand or manage it. Shop floor supervisors and managers are therefore pressured to increase direct labor productivity to save costs even though these costs are only a small fraction of the cost-of-goods, merely because 1) they are visible and manageable, and 2) they "drive" the larger overhead and therefore a disproportionate share of costs.

The ironic result is that, as labor content decreases and overhead costs remain the same (or increase), overhead rates must rise to completely absorb the same or higher costs on a smaller base. Therefore, investments in automation to reduce direct labor often yield disappointing overall results after the overheads have been reallocated.

Spreading overhead evenly on all products based on labor content is also not equitable. Any two products with the same direct labor content bear the same overhead burden without regard to other factors. If one of the parts is complex, it might involve more planning (more parts), handling, inspection, engineering, and fixtures, all of which are part of the overhead cost. A simple product with the same labor content, requiring few support services, will be unfairly charged the same overhead amount. Thus, the simple part will be overcosted and a complex part undercosted. Likewise, high-volume products, because of reduced direct-labor content, are typically subsidized by lower volume products.

Accountants, recognizing the dangers of the above situation, have been searching for alternative methods for defining and applying overhead. The objective of these efforts is to make overhead costs more visible and to have them more accurately reflect the activities that cause these costs to be incurred.

Activity Based Costing (ABC) is the current best candidate for a modern, more realistic overhead application method. Beyond that, ABC offers new opportunities to identify, track, and control costs that are buried in the amorphous "overhead" category in traditional costing systems.

The simple idea behind ABC is that activities consume resources (incurring costs) and products consume activities. By breaking the equation down into these two steps, the costs of activities can be more readily identified and the activities can more easily be tied directly to the products that utilize them.

Terminology

As is often the case when a new process is introduced, there is a unique vocabulary that goes along with it. Fortunately, there are only a few new terms associated with ABC, but they engender the basic concepts of the method in their definitions.

ABC costs are, in general, those indirect costs that we traditionally have called "overhead." Instead of lumping all indirect costs together into this one category and applying them crudely on all production, ABC encourages the creation of subgroupings of overhead called "pools" which can be more distinctly identified and more precisely applied. There can be as many pools as needed. A pool consists of all the costs associated with a cost category. Some examples of cost pools are: information systems, production space, quality inspection, the store room, shipping, maintenance, and many more.

The next step is to identify the "drivers" or consumers of the activities. The ultimate consumer of activities is the product, but we must be more specific and again subdivide those things that go into a product so that we can more closely associate costs with them. Examples of drivers include inventory transactions, amount of space required, number of orders handled (customer orders, purchase orders, manufacturing orders), number of projects, or number of units produced and so on. As you can see, drivers are mostly actions, and pools are categories of costs.

ABC comes to life when the actions and the cost categories come together. In a simple example, let us assume that you assemble a product that has five components, requires three operations on the plant floor (for a total of 2 hours direct labor), is inspected before packaging, and occupies 10 cubic feet of storage space when completed. Traditional costing for this product might look like this:

Material	$ 58.00
Direct Labor (2 × $12/hr)	24.00
Overhead (@ 320% of labor)	76.80
Total Product Cost	$158.80

Instead of applying overhead at the rate of 320% of direct labor, ABC would apply the various pools based on the directly associated activities (drivers).

Let us set up a pool for material (all inventory except finished goods). Included in the pool would be all costs associated with purchasing, receiving, managing the warehouse, material handling, incoming inspection, obsolescence losses, insurance, and MIS support. The total cost for all of this is calculated to be $800,000 per year, and records show that there are approximately 320,000 inventory transactions per year. The proportionate cost per transaction is therefore $2.50 each.

Our next pool is for production support. The total of all costs in this category is $1 million and we have a production budget of 100,000 direct labor hours. This driver therefore has an activity rate of $10 per hour. There is also a pool for tooling, space, handling, and maintenance that is applied at the rate of $40.00 per operation. Inspection cost is $8.80 per unit produced. Finished goods storage and handling is assessed at the rate of $0.70 per cubic foot, and there is a customer order charge of $14.00 to cover administration and shipping activities. As you can see, each pool of costs is spread evenly over the number of directly attributable activities to arrive at a "charge" rate per some measurable factor. Application of overhead can be much more equitable with the flexibility ABC provides to identify the most appropriate measurement for the driver (units, orders, hours, transactions, etc.).

If the product is produced in lots of 10, the cost breakdown for our example product looks like this:

Materials		$ 58.00
Direct Labor (2 × $12)		24.00
Overhead		
6 transactions @ 2.50	15.00	
3 operations @ $40.00/10	12.00	
2 hours labor @ 10.00	20.00	
Inspection @ 8.80	8.80	
10 ft storage @ 0.70	7.00	
1 order charge @ $14.00	14.00	
Total Overhead		$ 76.80
Total Product Cost:		$158.80

Although the total is the same, the ABC view allows us to see more clearly how the costs are incurred. Another product with the same direct labor content but with 12 components and 4 operations would have the same overhead cost under the old method, but would have an additional cost of $21.50 (7 additional inventory transactions at $2.50 plus one more operation at $4.00 based on the same lot size of 10) of overhead applied under ABC because ABC ties the cost to those activities that use them.

An interesting thing that happens when ABC is implemented is that people begin to see what support services cost, and they begin to question each and every charge. In the above example, inspection costs $8.80 per unit produced. The manager for this product might decide that he or she no longer wants to pay for inspection because quality has been improved to the point that there is no longer a need. The manager can now "fire" quality and thereby eliminate the cost of inspection. The manager of quality, on the other hand, has just lost a "customer" so he or she must now either find a new customer to replace the one just lost, raise the rates on other products to replace the income, or reduce costs to stay within the new lower "revenue" projection.

Increased management visibility and attention can often result in dramatic cost reductions after an ABC program has been implemented.

Contrast that with the traditional situation wherein the product manager, if he or she wants to reduce costs, has no choice but to focus on what he or she controls—direct labor. Rather than eliminating an unneeded service (quality inspection), he or she would have had to reduce labor content. The "fully loaded" labor rate is $50.40 per hour ($12 plus 320%). To save $8.80, labor would have to be reduced by nearly 10½ minutes per unit. Chances are good that this reduction will either require a significant capital investment or it may result in quality problems from cutting corners in the existing process. Rather than solve a problem, this change is likely to cause more.

Costs that have traditionally been viewed as "fixed" now turn out not to be fixed at all but merely invisible. Once the spotlight of the ABC approach is focused on the individual elements of cost and how they are applied (consumed), management can focus on true need, justification, correct applications, and reeducation.

Implementing ABC

Consulting companies, particularly accounting firms, see ABC as an opportunity to sell services; and for some companies, outside professional advice can be especially helpful in getting organized to implement ABC. There are also some software products available which are designed to help you gather and organize the information necessary to develop pools and track the application of costs. Neither of these is strictly necessary, however, to begin applying the principles of ABC.

As with most other endeavors, it makes sense to develop a plan to implement ABC in a controlled, manageable manner. Apply the "80–20 Rule." Seek out high payoff (high percentage of overhead expense) areas with only a few cost factors and/or easily identifiable drivers. Start simply and work up to a more complete implementation over time.

The first step is to identify candidates for cost pools. By looking at your overhead costs for the past year, isolate several significant cost areas that you can easily separate from the rest. Make sure that the identified areas can be associated readily to drivers which are already being measured or can be easily tracked.

Set the pool rates based on last year's actual expenditures, adjusted for any anticipated changes for this year, divided by a similar estimate of the driver volume for the current year. Note that some companies find that their general ledger account structure is not detailed enough to support the new breakdowns, and therefore some reworking of the ledger is required.

Next, recalculate overhead costs for major products using the new breakdown just developed. Work with these methods and assumptions as you apply the ABC principles and extend them to other cost pools. Based on the anticipated volume of production, establish a (monthly) budget for each pool and monitor actual application of costs against actual expenditures in much the same way as traditional overheads are reconciled. Make whatever adjustments are necessary.

Finally, be sure that the functional managers understand the purpose of this new accounting method, and receive regular reports of their products' (cost) performance. If cost reduction is a company direction, approach the challenge with an enhanced ability to see where costs really come from and focus on the activities that consume costs.

Don't feel that you have to break all overhead costs into pools and reapply them. Many companies have found that a few factors or categories account for a large percentage of the costs they want to track (the 80–20 rule again). Work with those costs that are significant and continue to apply the remaining costs in the traditional way. You can always extend your use of ABC at a later time, once you have gained control of the first pools of costs.

More importantly, be sure that the "users" understand what you are doing and why. Explain the concept and purpose of the exercise and be sure that the results can be understood and applied. Although there is significant benefit in being able to more correctly apply costs in your accounting systems, the far greater benefit of ABC is the increased management control and reduction of overhead costs that result from the use of the cost data by functional managers in the plant.

ABC in an MRP II System

Most conventional MRP II systems support only the traditional overhead application based on labor hours or labor cost. In order to apply costs per unit or per order or per transaction, some creativity must be applied. Several suggestions follow.

Per-unit costs can be structured into the bill-of-material as fictional component "items." Create an item for, say, the inspection charge. Use a standard cost and a quantity-per that applies the proper amount to the product. These fictional items can be coded as "floor stock" which requires no inventory transaction to apply the cost to production orders or schedules. You can also receive a quantity of this fictional item to stock that represents the budget for the month or year, and watch the balance on hand diminish as the item is "issued" to production activities. The remaining quantity on hand, plus or minus, represents the variance from the budgeted amount. Be sure that you can separate these fictional "items" from the real materials and components when calculating on-hand inventory value, preparing cycle count lists, etc.

Charges that are applied per manufacturing order or per operation can be included in the routing as set-up costs which are not applied based on order quantity or run time. Charges per transaction can be estimated for a typical production run and included using either the "item" technique or the routing approach.

This is by no means an exhaustive list of the ways to apply ABC costs in a conventional MRP II system, but it should be enough to get your creative juices flowing. The important point is to not let limitations of system design be a barrier to implementation of new ideas and techniques. Even if your application of ABC is tentative and limited, chances are good that you will benefit from a new perspective on product costs that can help focus management attention on those things that truly affect cost-of-goods.

A Real-Life Example

A small job shop manufactures marine exhaust systems as well as stainless steel boat hardware. The company suffers from poor profitability overall and uncompetitive pricing (costs) in the exhaust system product line. The table below illustrates their cost structure.

	Hardware			Exhaust Systems	
Material	$15.00			$150.00	
Labor (0.1 hr)	1.50		(10 hr)	150.00	
Overhead ($30/hr)	3.00			300.00	
Total C.O.G.	19.50			600.00	
Selling Price	25.00			750.00	
Gross Margin	5.50	(22%)		150.00	(20%)
Sales & Admin. (10%)	2.50			75.00	
Net Margin	3.00	(12%)		75.00	(10%)

To address the concerns listed above, this company separated the "indirect" costs of the foundry operation from those of the hardware fabrication shop. The result was a new overhead rate of $100 per hour in the shop and $15 per hour for the foundry. Recalculating the cost-of-goods and margin with these new rates, it became clear where the problems were.

It's no wonder that they were not selling many exhaust systems and overall profitability was poor. Management was now faced with a difficult problem: what to do about the hardware line. Obviously they could not continue to sell these products for $4.00 less than cost, but they also didn't know if they could raise the price and still get the business.

	Hardware		Exhaust Systems
Material	$15.00		$150.00
Labor (0.1 hr)	1.50	(10 hr)	150.00
Overhead (@ $100)	10.00	(@ $15)	150.00
Total C.O.G.	26.50		450.00
Selling Price	25.00		750.00
Gross Margin	<1.50>		300.00
Sales & Admin. (10%)	2.50		75.00
Net Margin	<4.00>		225.00

They screwed up their courage and talked to the customers, explaining the problem. The customers' response was that they suspected that something was seriously amiss because this company's competitors could not come close to their price and their quality wasn't as good either. As a result of these meetings, the company was able to raise their prices enough to restore the desired profit margin and retain their market share.

As far as the exhaust system line is concerned, they lowered the price from $750 to $650, enabling them to become competitive, while maintaining a healthy net margin of more than 20%.

In this example, a very simple approach was taken. Note that overhead is still applied on the basis of labor hours, but that two rates (pools) were set up based on the activities (drivers) that use them, namely, the foundry and the fabrication shop.

Effective use of the ideas of ABC need not involve high-priced consultants, M.B.A.'s, elaborate studies, or expensive software. The most important facet of ABC is to adopt the ideas of accountability and applicability. Identify the major costs in your overhead category, separate them out and identify the users, then tie the users directly to the service.

9. Computer Integrated Manufacturing (CIM)

A major trend in the evolution of manufacturing management systems is an ongoing attempt to more fully integrate (tie together) the business and management systems—those discussed in this book—to the other two major areas where computers are found in a manufacturing company: in engineering and on the plant floor. (See Figure 9-1.)

Computer Integrated Manufacturing (CIM) is the recognition that there are computers in use in these three distinct areas of the manufacturing business, and that the systems in each area contain and use information that can be of benefit if made available to the other areas of the business.

Historically, the application of computer technology to the needs of each of these areas has proceeded with virtually no recognition of the

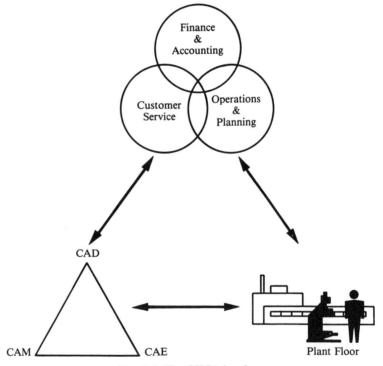

Fig. 9-1. The CIM triangle.

existence and needs of the other two. Therefore, there have been different vendors developing distinct, and for the most part incompatible, solutions in each area. Only recently has there been significant advancement in the ability to interconnect these various systems. In parallel with the development of hardware (and communications system) interconnection capabilities, software vendors are becoming involved in data exchange and the integration task to make use of the new connections.

The Three Areas of Automation

Integration within the business and planning area is well advanced with MRP II as well as the general office systems and decision support applications. These "front office" functions typically reside on a mid-range computer, although small systems (PC's) and networks have made some inroads in this area; and there are still a number of mainframe systems in larger companies assigned to these tasks. From a system standpoint, the business and planning tasks are primarily data management functions. The computers used here must be adept at multiple concurrent user access and be very good at storage and retrieval of large volumes of information.

As you know, MRP II handles the planning and management of materials, production, and purchasing. Production requirements and priorities are logically linked to plant-floor activities and should be electronically tied to the plant floor for efficient exchange of this information. MRP II functions rely on product and process definitions that are stored in its database. Much of these data are developed and maintained in the engineering systems, and there are some obvious efficiencies to be had by linking these functions together.

On the design side, Computer Aided Drafting and/or Design (CAD or CADD) systems have evolved down in price dramatically over the last few years. In the early 1980's and before, a high function CAD system required a mainframe computer and expensive displays and output devices (plotters) with an entry price in the hundreds of thousands of dollars. The breakthrough in CAD came as a result of the development of "workstations," powerful small computers capable of supporting the demanding tasks of handling mathematically described graphic images. While workstation computers are the dominant platform for CAD, the new more powerful PC systems are now in widespread use. The system task for CAD is raw mathematical calculation speed, called "number crunching."

Although networked workstations are the current platform-of-choice for CAD, mainframe CAD is still widely used, especially in larger companies with many CAD users. The large systems provide superior processing capabilities as well as large storage facilities and multiple user support for convenient access to a centralized "library" of drawings, standard images, and analytical functions. Once an object is mathematically defined in the CAD system, it can be analyzed using Computer Assisted Engineering (CAE) functions that have access to the drawing data file. These functions can display "solid"-looking representations of the object, simulate the motion of moving parts, allow clearance checking of moving parts, analyze the effect of stress and loads, and calculate the heat conduction and dissipation characteristics. All of these functions also require powerful mathematical calculation capabilities.

Also, based on the description of the object, a supplemental program can generate machine-tool instructions which can be passed to a production machine used to manufacture the object. The process of generating the instructions and using them to produce the object is called Computer Assisted Manufacturing (CAM). Note that some authorities consider CAM functions to be a part of the plant-floor group of applications.

Plant-floor automation evolved in the opposite direction from the other two areas. Rather than starting out using large computers then following the technology down to increasingly more powerful small systems, plant-floor computing started with simple machine controls and evolved upward. Early computerized production machines were given their instructions on paper tape and were known as Numerical Control (NC) machines. Fixtures and tools were set up manually, often requiring hours of preparation before running the programmed production sequence. Computerized Numeric Control (CNC) is associated with more flexible machines (more intricate programs and more memory) which are capable of multiaxis movement, more complex instruction sets, and sometimes automated tool changes.

The capabilities of CNC machines are limited in part by the amount of memory available for instructions and the time and effort required to load the program. These problems are overcome with Direct (or Distributed) Numeric Control—DNC. With these machines, the program instructions are stored on a separate computer which is not a part of the machine itself but is attached to the machine through a communications channel. The central storage and supervisory computer can be an advanced controller similar to the ones of the machines themselves, a PC, or a workstation.

Also on the plant floor, there is often a need to collect production information for quality monitoring purposes and to support the business and planning systems (job tracking, efficiency, cost accounting, payroll). Sometimes data capture equipment is attached directly to the production machine or could be built into the controller. Production information is most commonly collected through shop-floor terminals, bar-code data collection equipment, or hand-held terminals, if not manually recorded.

Machine controls operate in an environment that is quite different from that of computers that interact strictly with humans. In a human-based application, the system presents information on a screen and waits for the human to respond. When the operator completes his or her entry, the system's action then takes place in a burst of activity, whereupon the system's next response is displayed and another wait period ensues.

With machine controllers, there is a more continuous exchange with the other device. The controller issues an instruction, waits for a response (sometimes), then issues the next instruction. In many cases, the controller is monitoring a number of inputs (machine outputs) simultaneously and is comparing the input to a set of expected conditions. Based on the information received, the program can initiate various responses such as a command to move a cutter or reposition a workpiece.

These activities take place on a time scale in which humans can't directly participate. The communication between the machine and the controller typically happens many times each second. Programs for these applications deal with very small pieces of information, sometimes only a voltage level or a single pulse, and the instructions for the machine are often a single character or "word." Control computers are special-purpose microprocessors or microprocessor-based programmable units.

Integration

When a new product is conceived, that product might be drawn and specified on a CAD system. Using the CAD data, the properties of the new item can be tested for strength, movement, etc., with supplemental programs, probably on the same computer. Once the design is finalized, another program can create the machine tool program with which to fabricate the part. Item information and material specifications can be passed to the business and planning system to coordinate the acquisition of materials needed for production.

After industrial engineering has defined the process, and after planning has authorized production, the business system will assist in setting the production schedule and notify the plant floor as to how many to make and when. Materials are reserved, then issued with shop instructions and manufacturing begins. The machine program is downloaded from the engineering system at the appropriate time, and production information is gathered for quality assurance and to update the business system.

When completed, the part is moved to stock where it is available (just in time) for use in the assembly or product that needs it.

In most companies today, no such efficiencies are available. It is more typical to have the design prepared by hand or on a stand-alone CAD system. The testing of the part is done with a prototype, and revisions are manually added to the drawing. Part descriptions and material information are manually entered to the business system, and machine instructions are developed on the plant floor.

When it is time to produce the part, the program is retrieved from a file on tape or perhaps downloaded from a PC used exclusively as a program storage facility. Production information is recorded on time cards or reporting tickets keyed into the business system. Separate descriptions of the part exist in the CAD system, the machine instructions, and in the business system. Seldom are they coordinated when any of them is changed.

Past Directions

Historically, developments in the application of computer technology have advanced in the different areas of the manufacturing enterprise at different rates and using different technologies. In truth, the requirements of the three areas are different enough that there could probably not have been a general purpose processor powerful enough and flexible enough to satisfy all areas, even if there had been a strong motivation to design one.

In machine controls, the early task was merely to be able to accept a rather short list of instructions and execute them in sequence. No great intelligence was required, no data handling, no demanding input–output tasks; just ruggedness, reliability, and a very modest amount of storage to hold the instructions. As these systems have become more sophisticated, these basic requirements have expanded to process more instructions and have more ability to accept input from the machine and respond to these stimuli. More complex programs can be generated or downloaded, and many machine controls also include graphical output and programming-assistance screens as well as data-logging and communications capabilities. In comparison to today's business systems, these programming requirements are so fundamental as to be almost trivial, but this is not the strength of process control computers. Control computers must accept and process many inputs and outputs very quickly.

The response (cycle) time in these applications is measured in milliseconds or sometimes microseconds. When dealing with human operators, the difference between a half-second response and a one-second response is barely noticeable. In a machine controller, many functions can be performed in that one-half second.

Microprocessors constitute the dominant process controller computer type. These smart integrated circuits are relatively inexpensive, fast, and increasingly more capable. Microprocessors are teamed with memory modules to provide program storage and work space. Various devices to handle the input and output tasks, such as signal conditioners, analog-to-digital converters, line drivers, and display modules or interface devices, are used to manage the communications with the outside world. Often, all or most of these components can be mounted on a single circuit card for convenient packaging and common usage in a number of machines.

Business systems must be designed to handle lots of information efficiently. Storage requirements are very large, programs can be long but typically do not require extensive calculations. The primary tasks are to store and retrieve large amounts of data and be able to manipulate them: sort, summarize, and display. Business systems typically have a large number of interactive users (those having a two-way exchange with the machine) so these systems must be built to communicate effectively with multiple simultaneous users. Finally, business systems should include data management functions or software (called Data Base Management Systems or DBMS) to allow flexible access to the stored information.

Engineering systems are optimized for processing mathematical calculations quickly. The graphic systems used in the design and drawing functions really describe the lines and shapes of the images as mathematical equations. Manipulating these shapes, displaying different views (from different sides or angles), changing the size of an image or rotating it all require a tremendous number of calculations. For the system to be a responsive, these calculations must be performed quickly: millions of calculations per second.

CADD systems often are single-user oriented. Multiple-user CADD is usually organized as separate "seats" sharing a central file storage function but operating quite independently. Other engineering tasks, such as analyzing the figure or simulating motion, are equally or even more demanding from a calculation sense, so engineering computers, more than anything else, must be fast.

The vendors that develop and supply these systems have evolved along with the technology that was used in the areas they supported. Mainframe (and later mid-range) vendors developed and supported business applications; microprocessor technology grew up on the plant floor; engineering application vendors made their fortune in engineering. Inevitably, these companies tried to diversify into each others' areas, but generally were unable to displace the vendors who had nursed the technology (and the customers) along through the years.

As a result of the inability to "own" all three markets, vendors have gradually realized that the systems within the factory must be allowed to interact, despite their basic incompatibility. Thus, all manner of computer equipment vendors are beginning to recognize that they will not be able to displace all other vendors and provide for all of the users' needs, and that to survive with their piece of the business intact they must learn to work with the other vendors' systems. Every trade publication is filled with announcements that this company or that company has established a CIM division or has purchased an integration company or announced products that communicate with other vendors' equipment.

The time for CIM has arrived and it is none too soon. In today's increasingly competitive world, companies must take advantage of every facility available to increase efficiency and flexibility and to reduce costs and better control production. By taking better advantage of the computer power already in use in the plant, through integration, we hope to enhance our decision-making capabilities by being better informed of conditions and of the impacts of our decisions.

Integrating

The foregoing discussion notwithstanding, CIM is not just physically connecting together the various computers within the enterprise. CIM is an overall approach to management of the entire company and of the data and information that are part of the everyday operation of that company.

To implement CIM is to recognize that all areas of the company are interdependent. From initial concept through design, engineering, planning, production, customer service, and after-the-fact analysis, nothing happens without having an effect on what comes after or without being affected by what came before. To take a CIM approach to management is to recognize these relationships and to use them to advantage.

The most recognized aspect of this approach is known as Early Manufacturing Involvement (or Design for Manufacturability). As the name implies, it is getting the design department to listen to the people who will eventually be asked to produce the product. In this way, the design can be influenced at an early stage, where it is easiest to change, to make the product easier (more efficient) to build. Too often, engineering and manufacturing don't talk to each other except to point fingers and throw invectives during "How did we get this screwed up?" meetings.

Many times, a design can be modified slightly, without changing the function or outside appearance, to allow significant savings in manufacturing. This is especially important when automated processes are involved on the plant floor. While robots can do marvelous things, they are more limited in "manual" dexterity and flexibility than humans. A minor redesign of a component to include a chamfered edge, for instance, that is more forgiving of minor misalignment in assembly could increase a robot's success dramatically in day-to-day operation. While a design engineer might not consider this significant to the overall func-

COMPUTER INTEGRATED MANUFACTURING (CIM)

tionality of the product, a manufacturing engineer or technician might well recognize the benefits of the change immediately upon studying the drawing.

Putting the impact of a design change in perspective, think of the cost of a simple drawing change compared to a change after production has begun, which would include retooling, reprogramming machines, retesting (or certification), scrapping or remachining any on-hand inventory of parts, production disruption during the change process, documentation changes, etc., and you can begin to see the advantages of early involvement. Design for manufacturability has nothing to do with computer integration. It is the direct interaction of people from different business areas that makes it work.

There are some computer tools, of course, that can assist in this area. First on the list is Computer Aided Design that makes drawing modification so easy and efficient. Next are the Computer Aided Engineering tools that can be used to simulate the effect of changes, test the durability and other physical characteristics of the new design before a single chip flies, and can also simulate some of the production processes such as checking for possible interference as parts are assembled.

One of the computer-based tools that is becoming more important in design for manufacture is group technology. This is nothing more than a classification scheme that recognizes the similarities between parts. By grouping items together according to physical characteristics such as size and shape, and according to the production processes they employ, design can more readily recognize the factors that enhance efficient production and can more easily take advantage of past successes (and mistakes) in the design of new parts. Group technology can also help the designer recognize existing designs that can be reused or adapted to new applications, saving design and engineering time as well as reducing inventory and saving money.

Group technology is generally considered to be a tool for engineering but is most effective when used to make production more efficient. Should it then be considered production's responsibility? If engineering perceives group technology as something that they must do for production's sole benefit, where is their motivation to embrace it and make it work? And who is responsible for defining the coding scheme, assigning the codes to all items, and maintaining the integrity of the system?

These are not trivial questions. The success of any of the integrated systems or techniques discussed in this book is totally dependent on these kinds of considerations. Computers will do what they are programmed to do. People will do what they are motivated to do. Any implementation project must consider how the changes will be viewed by the people affected and whether or not the current environment will support the desired objectives. This especially includes the way people are motivated. As discussed in Chapter 11, incentives must directly encourage the desired behavior.

The points made about early involvement and group technology apply to all other interactions across functional lines. CIM requires that information be shared for the common good. It requires that each department and each employee feels that there is no reason to hold anything back and, indeed, feels a strong need to communicate freely and openly with the other areas of the company. The biggest impediment to implementing CIM can easily be interdepartmental rivalry or jealousy.

Often, this rivalry develops from competition for resources. Most companies exhibit a "personality" that favors the engineering side, the marketing side, or production. The bent of the company will reflect the key factor(s) in that company's past success. If the company is in a technical business, engineering is likely to be the driving force of the business. It is therefore also likely that high-ranking executives came from engineering and they are more likely to approve capital expenditures requested by the engineers than those requested by other departments. The same is true for companies dominated by marketing, production, finance, etc., as they tend to allocate resources more liberally in support of the area(s) that have been key to past success.

Whether or not there is justification for these feelings, the rivalry and competition for resources tend to discourage the free flow of information between functional areas. No department is motivated to make another look good at what may become, ultimately, its own expense. As new information bridges are erected in a CIM effort, it is important to recognize the proprietary feeling that people are likely to have toward their "own" information, and it is important to sell the benefits to each user, not only for themselves but also for the company as a whole. Fortunately, effectively implemented information management systems tend to reward the providers of information with a return that exceeds the value of the input. Through the synergistic effect of combining its own data with data from other sources into an integrated system, each function that contributes should receive back far more value than it could have received in a stand-alone application. This concept must be "sold" to the users.

Degrees of Integration

System integration can take many forms—from very simple exchange of information to full interaction between applications. In the simplest case, data exchange, one application stores information which another application can access. If both applications are on the same computer, the exchange is relatively straightforward. This form of integration is the common thread in most application families such as MRP II and CAD/CAM.

When the applications are on different systems, the data must be provided with transportation between the systems and there might be duplication involved. The simplest exchange of this type is to copy a file from one system, carry over to the other system, and read it in. Sometimes jokingly called "sneaker net," the physical transfer of data via diskette or tape is the least efficient and the most dangerous type of exchange. Once information has been duplicated, you must be concerned about updating and synchronization.

A better variation of the same idea is to have the application on one system directly access the data on the other. If the systems are alike but separate (two PC's, two DEC VAX systems, two Sun workstations), the primary requirements are physical connection via communications lines, and applications or utilities that are able to use the connection. This avoids duplication of data but can be hampered by the speed of the communications link, by security concerns, or by differences in applications or operating systems. If the systems are different, translation facilities are also required to make the data useable. Dissimilar systems can be connected through emulation utilities (to make one system look to the other system like an ordinary terminal or other device), or protocol converters. Similar networks are connected through "bridges," while dissimilar networks are linked via "gateways."

The second level of integration involves a true sharing of responsibility for the information. More than one application cannot only access the data but can also maintain and update them. Again, this is common practice in MRP II where a centralized database is shared by all applications in the set. Each function is authorized to not only use but also to update data as required within the range of its duties and responsibilities. True data sharing is difficult and seldom used outside of a single system or network.

The highest level of integration is interoperation of applications. Also called program-to-program interaction, this is an interactive exchange between applications rather than store-and-read exchange of data. Interoperation is often seen today in implementations of cooperative processing (CP). A typical CP application links a PC-based data manipulation utility to a mid-range system application.

Since today's PC's are very good graphical user interface platforms, while mid-range systems have excellent data management capabilities but are pretty much strictly text-oriented, the combination of the two can provide the best of both worlds—mid-range power and PC user-friendliness. Microsoft Windows and IBM OS/2 applications with colorful point-and-click displays are being used to "front-end" database utilities to conveniently access, summarize, and present selected information from the larger systems.

Integration Goals

There are a few characteristics of integrated systems that you should look for in assessing the degree of integration. These few points are simple but contain the essence of the idea of integration.

* *Data are in only one place.* A particular element of data should not be duplicated if at all possible. Whenever there is duplication, there must be a concern for keeping the two versions of the data in "synch," that is, making sure that both are updated at the same time. As the old saying goes: "Show me a man with a watch and I'll show you a man who knows the time. Show me a man with two watches and I'll show you a man who's not sure."

There is a common problem with PC's that illustrates this principle. Many companies have numerous PC's throughout the facility, most of them with spreadsheet and database packages installed. It is common practice for managers to create spreadsheets to analyze and manipulate data from the corporate system database. As soon as corporate data are loaded into a PC spreadsheet, they become obsolete. When these data are manipulated in the spreadsheet, they diverge even further from the centralized source data which are themselves changing as activities are reported. I have been in meetings where heated arguments have broken out over which spreadsheet is most up to date. "Show me a man..."

* *Data are entered once.* The same concern applies to transaction entry—a single entry should update all affected data. When applications are designed to be integrated, using a shared database, this requirement is usually easily satisfied. When add-on applications are developed, or integration is done after the fact, there is often duplicate entry required.

* *Single device access to all data.* Systems should be tied together such that only one terminal device is required to access all applications. Further, conformance to design guidelines such as IBM's Systems Application Architecture provides the user with a consistent interface format, i.e., screen organization, use of function keys, application flow, to reduce confusion and the need for extensive retraining for each application.

CIM Facilities

A relatively new entrant in the field of manufacturing support hardware and software is the CIM System. IBM started the ball rolling with the announcement in late 1989 of their "CIM Architecture" which is a design and overall structure for a CIM facility to be used to draw together the three areas of automation.

CIM Architecture, and similar product ideas from other vendors, provides a set of central utility services and interfaces designed around the concept of centralized data management and coordination. The system provides a storage facility for common data and a directory service for all data whether common or unique to a specific application.

When fully implemented, an interfacing application could request data from any application in the system without having to worry about location or format. The request is handled by the directory service, called the "Repository" in the IBM product, which determines the location of the requested data. If the data are in the central storage area ("Data Store" in IBM), they are passed to the application. If another application has control, the repository service accesses the data and transfers them to the requesting application.

All of this is automatic and invisible to the requesting applications. The CIM Architecture takes care of all the details in much the same way as a Data Base Management System relieves the application program from the tasks of data definition and specification.

A CIM utility such as this cannot simply be installed with existing applications. The applications themselves must be rewritten to relinquish control of common data (that which is shared with other applications) and taught to ask the repository for data rather than being limited to its own database.

IBM's market presence has enabled it to instantly establish a de facto standard for CIM services. The largest vendors of MRP II on IBM's midrange computer system (which represents nearly half of the MRP II installed base) have already developed interfaces to the IBM CIM Architecture, as have the major CAD system vendors (on IBM platforms). IBM also offers numerous "enabler" products to be used to bridge to plant-floor devices and PC-based applications.

Why CIM?

In most products and markets today, the primary competitive factor is, or will soon be, time—the time it takes to develop and release a new product, the time required between the receipt of a customer order and shipment of the product, the time it takes to react to changes in demand patterns. CIM offers opportunities to save time by eliminating delays in getting information from one application to another and in enhancing the ability of the various departments to work together through a new feeling of cooperation and joint interest. The need to cooperate for successful system implementation tends to break down barriers that exist within most organizations.

By enhancing the flow of information through the company, CIM can help reduce the delays and needlessly complex processes that can hamper a company's responsiveness.

CIM can also help a company improve quality in several ways. Because there is less confusion and more coordination with CIM, it is more likely that products will be designed for easier manufacturing, will be handled less and tracked more closely, and will probably suffer less from expediting and other harmful practices.

CIM facilities will also help provide a communications system throughout the enterprise to facilitate control and tighter management for more efficiency and responsiveness, particularly in support of continuous improvement programs. It's difficult to improve what you can't measure, and difficult to control what you can't track.

10. Software: Selection, Adaptation, Modification

It is generally recognized that there is no such thing as a perfect fit of packaged software to a company's specific needs. Given the complicated nature and the diversity found in manufacturing, most would find any package that satisfies more than 80% of the requirements to be a good solution. Does packaged software make sense, then? Absolutely.

Packaged software offers many advantages over custom-written software, not the least of which are low cost, proven function, availability of education and support services, and ready availability. To develop your own custom applications today is almost always too expensive, takes too long, and offers significantly less capability than a packaged solution, despite the less-than-perfect fit.

Ironically, custom software is hardly ever a perfect fit either. Because of the practical limitations of time, budget, and capability, custom applications cannot be as complete nor comprehensive as a packaged product. During the extended development cycle, the company's needs will change, and the design of the application will either evolve with the needs or be outpaced by them. If the software specifications evolve during development, the initial design will almost certainly prove inadequate, and either the function will be compromised because of these limitations or costly redesign will delay the project further, making the suitability of the design subject to further erosion.

Changing specifications are a software developer's nightmare. This is the most frequent cause of dissatisfaction and rancor between the developer and the customer (user), whether the development is being done by an outside contractor or your own data processing staff.

What are the options, then, when faced with the prospect of selecting and implementing a less-than-perfect package? The first challenge is to find the package with the best fit. Then, either learn to live within the limitations of the package, adapt the standard offering to better serve your needs, or you may choose to modify it.

Selecting a Packaged Solution

There are literally hundreds of packaged MRP II solutions available today. They range in price from hundreds of dollars to more than a

132 SOFTWARE: SELECTION, ADAPTATION, MODIFICATION

million, and are marketed by computer hardware companies, general software companies that have packages for many applications, manufacturing software specialists, systems integration companies, accounting firms, consulting firms, the computer store at the mall, and the whiz-kid down the block. Choosing the right package can be a real challenge.

The Request for Proposals

To find the best fit, you must first understand your needs. The discussions in this book should help you identify the primary areas in which companies have found basic MRP II to be lacking, and the kinds of features and solutions offered by software packages to address these needs.

Some companies, especially larger ones with deep pockets, will commission an outside consulting firm or accounting firm to prepare a formal request for proposals (RFP) which can be sent to a number of vendors, soliciting formal presentation of their solutions. Having been on both sides of this approach—as a consultant preparing RFP's and as a vendor responding to them—I have mixed feelings about the validity of the RFP approach. On the one hand, it is essential to analyze and document your needs, and communicating these needs effectively to the vendors gives them a good opportunity to respond with a valid proposal. On the other hand, however, this process is often overdone. I've seen companies spend more on the analysis and RFP process than it would have cost to buy and install a very good mid-range solution.

Some of the larger accounting firms will charge tens of thousands of dollars for this service, and will prepare a document that is made up mostly of "canned" questions that may or may not focus on the critical issues. Often these "professional" RFP's will contain hundreds of questions, most of which describe in minute details the features and functions that virtually every MRP II system worthy of the name will include as a matter of course. Worse yet, these questions are often what the vendors call "wired," that is, they describe the details of one particular package, therefore making it difficult for any other package to look good by virtue of having too many "no" responses.

This "wired" nature is sometimes inadvertent, as a consultant who is most familiar with one particular package will include its features as minimum requirements and often will not know details about features of other packages that may be equally or more important for this company. The problem is magnified when the preparer firm is associated with the sale and/or installation of a package of its own.

The problem on the respondent's end is that these extremely detailed RFP's are often received with very short deadlines for response. I have on

SOFTWARE: SELECTION, ADAPTATION, MODIFICATION 133

many occasions been asked to respond to a questionnaire of more than one-hundred pages, with over a thousand detailed questions, in just one or two days. The RFP may well have been sent out with a month to respond, but the delay in delivery, routing to the proper sales rep, initial assessment (do we want to respond?), assembling the right technical team, etc., can easily use up most of the time.

When there are a thousand questions, the bidder can't afford to do more than a cursory analysis, and the responses are often limited to: 1) we comply with this requirement, 2) we don't do it quite this way but we do something similar, 3) it's not a standard feature but we can adapt (modify), or 4) no, we don't have this. Writing the detailed explanations for answers 2 and 3 is a challenge, especially when done on a very short deadline. In addition, there is limited opportunity for the vendor to promote the unique advantages of his or her product within the relatively rigid framework of the RFP questions.

The evaluation of these proposals is often done with a point system for scoring. The number of positive responses, or qualified yes's, and no's, are tallied, sometimes weighted to reflect the importance of the feature. There is little room for judgment within this format—the item is usually either satisfied or not. Total weighted scores are used to reduce the number of contenders to a reasonable number, usually two to five. The finalists are then scheduled for presentations and demonstrations before a final selection is made. The entire process takes a minimum of three to four months, often longer.

The RFP can, however, be made to be a useful approach to system selection without being overly complex or expensive. Rather than hundreds of pages of marginally useful, detailed "boiler plate" questions, the RFP format could be used to describe the business needs of the company and focus on the primary areas of concern. It should describe the current situation, primary concerns (problems) and the goals and direction of the company for the next three to five years. It should invite the bidder to visit the plant and talk to the prospective users, all the better to understand the company and its situation.

In addition to functions and features of the software, the RFP should seek information about support services and implementation assistance. What do the bidders have to offer, and what other sources are there for support?

You will also want to know about the vendor(s) you will be dealing with. How big are they? How long have they been in business, and how stable are they? Insist on references and be sure to talk to some of them. Try to find satisfied users of similar size with similar needs.

Just because the vendor company is small should not automatically eliminate them from the competition. You must be careful, however, to protect yourself against the possible demise of the vendor. With smaller companies, check references very carefully, insist on the use of widely supported hardware and standard computer languages and data utilities. Also ensure that you have access to source code so that you can find other support for your system if the company disappears.

The Invitation to Bid

A less formal approach involves the identification of a select number of vendors who are invited to visit your plant and gather their own requirements inventory. The initial selection might include companies with a strong local presence, those whose software runs on the equipment you own or are familiar with, or the largest vendors in your industry segment or hardware size category (mainframe, mid-range, or micro). You will spend more time ushering the potential bidders around the shop but you will save the up-front effort of doing the analysis yourself and writing the RFP. Sometimes several bidders can be accommodated with a group presentation and tour followed by private question-and-answer sessions.

The responses (proposals) thus generated will vary more than the answers to RFP questions, and the comparison and selection will probably be more subjective, but the proposals might also make you aware of problems or opportunities that you might not have discovered on your own.

The Sales Call

The solution may come, unsolicited, to your door. Sales reps still make "cold calls"—through the mail, on the phone, and sometimes door-to-door. Marketing "seminars," demonstrations, and other events are held regularly to interest prospects in the products. These events are often very educational and don't usually involve high-pressure sales techniques. If you see something you like, invite the vendor to tour your facility and propose a solution.

I don't necessarily advocate that you buy a system based on one proposal from one vendor, but it's not necessarily a bad way to do business. If you have a vendor now that you are satisfied with, why not let them propose the next generation system for you? If the vendor that sought you out has a good reputation, an appropriate solution, and you feel comfortable with them, why not? As you have seen, there are accepted approaches to the big issues incorporated into many off-the-shelf packaged solutions. If the bidder's package has the right features and meets all of the other criteria, there might not be any need to look further.

SOFTWARE: SELECTION, ADAPTATION, MODIFICATION   135

So, You've Decided

Before moving on to the discussion of how to handle the unique requirements—that last 10% or 20% of the "fit"—I'd like to add a few more words of caution.

There is a lot of turnover in the software business. If you know how to program, or think you do, the only thing you need to go into business is a phone and some business cards. Every day, new software companies start in business and others disappear. Most software companies don't last. Even the big ones go broke, merge out of existence, split up, file for bankruptcy, lose their best people, discontinue products, and otherwise become less capable of supporting customers. If your business depends on the continued operation of their software, you are vulnerable.

All software has bugs. Some of them don't show up for years. As your business grows and changes, you may begin to use parts of the package that you never used before. That error was there all along, you just didn't know it. If the vendor is gone, how will you solve the problem?

There are several ways to protect yourself. First, be sure that you have access to the source code for the programs in your package. Most vendors will sell (a license for) the source code at a modest additional charge, or sometimes will include source at no extra charge. If your selected vendor refuses, seriously consider another vendor. An alternative is to have the source held by a third party with a contract that gives you access to the source if the vendor is unable or unwilling to provide support. This is a much less desirable solution, but is one way to protect yourself if you are not allowed free access to the source code.

Another safety valve is to buy a package that is supported by third-party resources, that is, someone in addition to the software vendor. It is good to have available several programming companies, authorized representatives, systems integrators, or other firms, independently owned and managed, who know the package and can help you support it in the event of the vendor's demise or if you decide that the vendor is not supporting you to your satisfaction. You hope it will never happen—you select your vendor carefully to help ensure that—but you can never know for sure.

Learning to Live with Limitations

It is a given that no packaged product is a perfect fit. Hopefully the differences between your needs and the package's capabilities are minor, and you can use the package, despite its flaws, pretty much "as is." In the popular jargon, this is known as using the package in "vanilla" condition.

If you have made an appropriate selection, the package should provide for the majority of your company's needs. There can sometimes be some differences in terminology or format between the new package and the old way of doing things, but it is highly recommended that this kind of difference be accepted and you should adapt your procedures to the package's way of doing things rather than modify the package to look like the old way (more about modifications later).

In addition to the considerations outlined below, a danger in changing a package is similar to a problem with custom development: you will design for today's needs and sometimes preclude enhanced function that you don't need today but may be able to exploit in the future. The old way of doing things obviously had some disadvantages—otherwise, there was no reason to buy the new system. If you replicate the old way with the new system, what have you gained?

Since a packaged product must be designed for a wide range of situations, the standard reports and inquiry screens will tend to be "busy," that is, will contain more information than any one user will need, in order to accommodate as many situations as possible. New users can find this annoying or confusing. Resist the temptation to remove unneeded data, if you can. It is often the case that some of those "extra" fields really are beneficial; but if they are removed, you might never uncover their usefulness.

I have seen many cases where a company has modified a package to include a "new" function when the desired capability was always there but was hidden by early modifications that removed information and capabilities that the company either did not need at the time or did not understand enough to appreciate them. Under the next section, I will discuss additional reports and other add-on functions that might have the same effect as changing reports or inquiry screens. Be careful. The short-term convenience might have long-term detrimental consequences.

"Standard" Functions

Virtually every company that I have worked with over the last twelve years has thought themselves unique. The truth is, however, that most manufacturing companies do similar things in similar ways and the uniqueness tends to be more one of terminology, technique, or details. Since packaged products are designed to meet the needs of a large industry segment (all manufacturing, all process manufacturing), the functions and procedures included are pretty much the "industry standard" way of doing business. Before modifying a package, you should ask yourself why you do whatever it is you do in a way that is significantly different from the way other companies do it, as incorporated in the "vanilla" package.

I understand that companies like to feel that they have a competitive advantage because they *are* different, but often the differences we are talking about here are not significant in terms of a competitive edge. Usually they are the result of long-standing tradition, inertia, or ego (the "not invented here" syndrome). Once you have identified the "why" behind your way of doing things, decide if your uniqueness is really a significant advantage—significant enough to justify the trouble and expense of modifying the packaged software.

Implementation experts will talk about modifying your procedures to adapt to the package rather than modifying the package to adapt to your procedures. This is by far the preferred approach for all the reasons discussed below (modifying a package) and is usually the fastest, most rewarding approach you can take to implementation. Users will resist changing their ways. There is a natural human resistance to change. This resistance can be overcome by adequate education in the functions and capabilities of the new package and through an understanding by the users of the advantages of the new approach. The latter point is achieved through education, and via firm direction and commitment from above. Above all, remember that each user will approach the proposed change in his or her process or procedures with an attitude of "what's in it for me?"—although few will actually voice this concern so directly. In order to wholeheartedly accept and support a change, a user must feel comfortable that there is adequate justification (reward) to compensate for the inconvenience, conversion, retraining, or whatever adjustments are needed to put the new function into use.

Adapting a Package

If the package is deemed to be inadequate in some significant area to the extent that a change is required, there are several degrees of change, with varying impact on the maintenance and supportability of the package.

I like to distinguish between "enhancements" and "modifications." An enhancement can be custom developed or sometimes can be purchased from the original vendor or a third-party software supplier. An enhancement is either totally or mostly "outside" of the package. By way of contrast, a modification is a change that directly impacts the purchased program(s). This section describes enhancements. The following section deals with modifications.

Typical enhancements provide additional reports, add auxiliary functions (enhanced sales analysis, reformatted control documents, regulatory compliance reports, enhanced inquiry functions, industry-specific analyses), or extend the function of the system into niche industry segments. Modifications often change the logic of a process such as new or enhanced cost accounting functions, unique industry-specific calculations, or changes to input formats or data requirements.

If the package you chose is widely installed, chances are good that companies other than the vendor have developed additions to the package and market them as add-on enhancements. The existence of a rich supply of enhancements is not necessarily an indication of inadequate function of the base package, but more likely reflects the diversity of the marketplace and sufficient installed base to justify development and marketing of add-ons.

Packaged enhancements offer many of the advantages of packaged software as discussed earlier: packages are much less expensive than custom developed code, they are supported by the vendor and are available off-the-shelf with no waiting for the development to progress. As with any other software purchase, check references, satisfy yourself that the vendor is reputable, verify that the function purchased is what you need, and protect yourself with access to source code.

Although many packaged enhancements are completely outside of the base package, that is, they do not include any modifications to the primary package programs, there are some enhancements that do touch the base code. Be careful here, as the line between enhancement and modification might become a bit blurred. When the change affects the basic package, be especially careful. Assure yourself that the modification to the code was done carefully, is fully documented and supported, and was designed for minimum interference with the base function.

Support is critical. Most software vendors provide fixes and updates on a regular basis to registered users. Any time there is a change made to the vendors programs, the update process becomes problematic. You bought the basic package from vendor A. You bought an enhancement from vendor B that modifies several of vendor A's programs. Vendor A sends you an update. You must now contact vendor B to find out if their enhancement is impacted by vendor A's update. If so, you must wait until vendor B sends you *their* update before installing vendor A's update. Install A's update first, then B's update, then test to be sure that both vendors did things right. If there is a problem, call vendor B first, try to eliminate them as the cause of the problem before calling vendor A. All of this obviously takes time and effort.

If you write your own enhancements, the situation is similar, but in this case vendor B is you. When you receive an update from the primary vendor, you must check the function of your enhancement with the new code, make any necessary changes, and delay the implementation of the update until you have made and fully tested the necessary changes to your add-on.

Of course, you always have the option to forego the updates offered by the vendor, remaining at a particular level of the programs without installing additional updates. This state is known as "down-level" and pretty much eliminates any possibility of vendor support if you have any problems. This may be a viable option in some circumstances, but most companies, sooner or later, regret a decision to stay down-level. Eventually your business will change, your market will take a sharp right turn, or new management will want to take advantage of changes in technology, and you will be faced with a difficult decision and an arduous path to migrate to a new solution or the current version of your chosen package.

Modifying Packaged Software

It is perhaps unfortunate (but it is true) that many packaged software users choose to modify their packages. Why? Many times it is the result of a firm conviction (belief), right or wrong, that their business is unique enough that the standard functions are inadequate. Whether justified or not, the users have demanded changes or additions and the MIS staff (or senior management) has opted to accommodate the requests.

The support and upgrade considerations for modified software are the same as with enhancements but to a greater degree. A modification as I have defined it is a change to the logic of the vendor's programs and thus will be impacted by any update that touches those same programs, and likely will have impacts beyond the individual changed programs.

MRP II systems are, by their nature, integrated. Data are shared. Functions are interdependent. Changes in one area almost always affect one or more other areas within the system. When designing and implementing changes, it is extremely important to thoroughly test for these interactions and understand how data flow within the system.

It is for this reason, as well as the others discussed throughout this chapter, that modifications tend to be more expensive than anticipated and more difficult to install and support.

Controlling the Change Process

Too often, the MIS department is given the responsibility for deciding what modifications are to be made and which requests are rejected. This can put MIS in a very difficult situation. MIS, being a customer-oriented service organization, will want to be responsive to its customer—the user. If they refuse to make a requested modification, they can be viewed as unresponsive or uncooperative.

To make matters worse, MIS often reports to the head of another functional department such as finance or administration. If there are thirty requests pending and only enough resource to satisfy twenty, which department do you suppose will get more approvals?

It's really a no-win situation for MIS in such circumstances. They might fully understand the need to limit modifications, but there is considerable pressure within the organizational structure to modify. Their measurements are probably tied to the number of projects completed or the amount of code written. To refuse or limit modifications would work against their measurement system.

My recommendation is to take the basic decisions out of the hands of MIS and give them to a high-level committee with representatives from all major areas of the company. Such a review committee would not be unduly influenced by any particular department and could look more objectively at user requests for modification.

The change control process should be formalized, with every request accompanied by a justification. MIS would obviously be involved in helping the requesting users identify the extent of the change requested and the cost, whether performed in-house, by a contractor, or satisfied by a packaged enhancement. Cost figures should include long-term support considerations. The user would also be required to specify the benefits to be achieved as a result of the change. If the benefits don't outweigh the costs, the committee wouldn't even have to see the request—it would self-destruct in the justification process.

Those requests that make it to the review committee should then be carefully considered, accepted or rejected, and the accepted requests prioritized and scheduled. As a result, MIS has only clearly defined, justified, and preapproved projects and also has the benefits of an unbiased oversight group to resolve conflicts and take the direct pressure of user demands.

Minimizing Impact

There are several significant advantages to keeping your package as "vanilla" as possible. Although some vendors will attempt to support a user who has modified his or her package, others will simply refuse.

Even if the vendor is willing to try, it is quite difficult to support code that is a combination of standard product and changes instituted by the individual user company. Let's say that your shop order release subsystem fails when attempting to initiate an order for an unusually large quantity. If you have made changes to any of the programs involved in the release process, you will not know, at first, whether the failure is a result of your modification or a failure of the vendor's product.

If you call the vendor's support line, one of the first questions they are likely to ask is whether there are modifications present. If the answer is yes, and the vendor is still willing to help, you (someone) must identify whether the error occurs in unmodified code. Either the user company or the vendor will try to duplicate the conditions in which the error occurred—not an easy task—and isolate the cause.

If the fault is the vendor's, after they correct the error it will be necessary for you to reapply your modifications after the corrected programs are provided by the vendor. If the fault is yours, you will be embarrassed and the vendor will be annoyed, at the very least.

The best modification is the least modification. When you are aware of the consequences of code changes, you will work to minimize the changes to vendor's code. There are a number of techniques that can be used to achieve this goal. Often, several of the following techniques can be used in combination.

* Use separate programs and subroutines for new logic. Insert only a line or two in the vendor's program to "exit" to your function and return when completed.

* Copy the vendor's program, insert the change, and retain the original program. This allows you to revert to the original for testing and problem determination. Some systems use a "library list" organization that allows you to have multiple versions of a program on the system. The modified version is put into a list with higher priority than the original version. In this way, the vendor's code remains "pure" in the vendor's library and all modified programs are together in a higher priority "mod" library where they can be easily managed.

* Fully document all changes. Documentation cannot be overemphasized and is usually missing, incomplete, or poorly managed. Be sure that the effort required for documentation is included in the costs used in the change justification, and be sure that it is done. I suggest, at a minimum, internal documentation (program comments), lists of modified programs with description of the changes, annotated program listings with the change highlighted, full documentation of any database changes, updated user documentation (user manuals, procedures), and a log of support issues and maintenance activities.

* When you remove a line of vendor's code, do not delete it but inactivate it by changing it to a comment. Every programmer knows how to "comment out" a line of code.

* Avoid changes to the vendor's database. Even with relational database management systems, keep your changes outside of the vendor's architecture as much as possible. Add a new file or reuse an unused field rather than add a new field to an existing file.

Summary

Even with the best advice, a top-notch systems staff, and a well-executed change management process, the best strategy is to pick a package that provides most of the function you need, adapt your procedures as much as possible in the areas where there are differences, and minimize any changes to the vendor's code if you must modify. The main reasons for buying a package are: to be able to take advantage of superior function faster and at less cost than custom programs, to benefit from support services that the vendor and third parties can provide, and to be able to keep up with the technology through the update process. All of these benefits are jeopardized by modifications.

Remember that no package is a perfect fit. Most companies find that careful selection will result in about 80% of the requirements being well satisfied by packaged functions. One key to success is how you choose to handle the other 20%. The 80–20 rule works both in your favor and against you. The package can provide 80% of the function (or more) at 20% of the cost (or less), but modifications can supply only a small fraction of the function at a large percentage of the cost.

I have seen many cases of a one-hundred-thousand dollar package with a half-million dollars in modifications (and equivalent spending in other price ranges). And that is to say nothing of the implementation delays caused by modification, the inability to use standard education and documentation, and the support costs and difficulties.

11. Measurements

When you go to the company's senior management or board of directors and ask for a rather substantial chunk of money for an information systems project, they are likely to ask what the return will be. To get the allocation, you will probably have to convince them that the benefits gained will exceed the investment made by an acceptable margin. If you don't have the answers already prepared, you haven't done your homework. If they don't ask for this justification, they aren't doing their job.

Even if you ignore the issue of investment/return, there is probably nothing more vital to the long-term health (or maybe even survival) of your business than effective management of information. It seems only prudent, then, to talk about the impact of MRP II and how to measure the effectiveness of your implementation.

With any measurement, two or more data points are much more useful and informative than one. Not only will subsequent measurements validate the initial reading, but multiple measurements over time will indicate whether the situation is improving, degrading, or not changing at all.

When preparing to begin any kind of improvement project, it is important to know what your starting point is. Many (I would have to say most) systems projects fail to establish baseline measurements against which progress can be measured. A year or two down the road, when they are ready to go back to the well for the next project, it can be really embarrassing to not be able to quantify the payback from the previous project. Put yourself in the following situation.

You requested, and obtained, one-million dollars to implement a basic MRP system. After eighteen months, you go back to the board for another quarter-million to add data collection equipment and several additional software modules. One board member, who happens to also be the company controller, asks if you actually got the return you promised for last year's million. You say that the modules were installed on time, the users have taken to the new function, things are running smoothly, you got production to report their activities every day instead of once a week, and so forth.

The controller is getting impatient, you can see it in his or her eyes. You start talking a little faster. The controller starts fidgeting. Finally he or she stops you and says, "How much have we saved as a result of this project? How much have we reduced inventory?"

"This report," the controller says, pulling a printout from a stack on the table, a triumphant look barely hidden behind horn-rimmed glasses, "says that inventory is up four percent."

"Er, ah...well..." You just lost your data collection system.

Had you been prepared, you might have been able to explain to the controller that the increase in inventory was in support of an increase in business volume and was, in fact, a reduction when viewed as a percentage of sales. You could have shown the trend of inventory-to-sales for the past year (a steady downslope) and pointed out that component inventory, that which is of primary interest to MRP, is in fact at its lowest level ever (relative to sales). You might have also been able to show that finished goods inventory is up only slightly (in dollars, down as a percentage) but customer service has improved dramatically, which is part of the reason why sales are so strong despite the weak economy...

You would not be able to make your case if you had no measurements of what the situation was before you started implementation, as well as periodic measurements since. In truth, most MRP II users have few, if any, measurements and those that they have are often not very informative.

The primary purpose of measurements, however, is not financial justification of the next project. Measurements should be tools that help the system users judge the effectiveness of what they are doing, identify opportunities for improvement, and measure progress toward established goals. These kinds of measurements are typically not financially oriented. More often, operational measurements are based on time (dates) and quantity. When effective operational measurements are in place, however, they can usually be "dollarized" readily to provide the financial measurements that top management will probably want to see.

The most important facet of a measurement is that it must relate to the goal for the process being measured. In line with appropriateness, it must be useable, that is, understandable by the person(s) doing the measurements. To meet these two criteria, it must be specific (each work center, not the whole plant), relevant (units scrapped or hours spent on rework, not total rework cost), and must be an integral part of the improvement strategy.

A classic situation is the measurement of production in terms of dollars of product completed per month. First, production doesn't make dollars. They make "each's," pounds, or gallons. Second, a measurement such as dollars per month ignores priorities, shipment schedules, need, and appropriateness. With an MRP II system, production schedules are tied directly to shipment of products to customers (to the Master Schedule, in other words). If the company goal is to ship on time and make the best use of resources to do so, then the measurement should be tied to on-time production completion or adherence to system-generated priorities.

Other common measurements that are no longer relevant in many cases are efficiency and/or utilization. Since direct labor is now such a small part of cost-of-goods in most industries, efficiency needlessly emphasizes an insignificant factor (see the discussions in Chapter 8). Measures of efficiency motivate the production workers to focus on "beating the standard" at the possible expense of quality and without regard to priority (tied to customer service goals).

Utilization (keeping resources busy) tends to build inventory. A supervisor who is measured on pieces produced or amount of idle time is motivated to produce more parts than needed (longer runs use more of the day for production), to advance jobs ahead of schedule to keep the number up ("That other job runs easy so I'll do it next; I'll let the next shift worry about this "hot" job that's waiting"), and to make unneeded parts just to keep busy. We have to learn that the old adage "a busy machine is a happy machine" just doesn't apply any more. It is far better to have idle machines than to produce parts you don't need.

Basic Measurements

VENDOR PERFORMANCE

What is more important to the business: vendors who deliver quality products on time, or vendors who might be a little less expensive but are not as reliable? The answer is obvious, but we still tend to measure our buyers on price variance.

Many of today's MRP II systems contain the capability to monitor a number of vendor (purchasing) performance factors including early and late delivery, meeting specified lead times, over/undershipments, and quality as well as price. It is important to change the way the people in the purchasing area, both internally and at the vendors, are measured and motivated.

Some companies, when they institute a new vendor performance system, will send a letter, visit the vendors, or invite the vendors in to discuss how their performance will be judged in the future. If the vendor doesn't know what you consider to be important, how can they respond?

Just be sure, when you do this, that the performance factors that you designate as the most important are the ones that directly support your company objectives.

ON-TIME PRODUCTION

Assuming your master schedule reflects a realistic plan to satisfy customer demands, your primary concern should be whether production activities are completed when due, just as you want to ensure that purchase receipts are on time. Provide the plant floor with clear direction in the form of valid due dates (that don't change daily) and priorities, and find ways to measure conformance to these priorities.

On-time completions are certainly a measure that you will want to watch, but not to the exclusion of other considerations. Most plants, no matter how out of control, manage to turn in a fairly good on-time shipment record. Unfortunately, many do it through expediting, overtime, constant turmoil, and great unnecessary expense.

One thing to consider is whether the plant is given sufficient lead time to complete the jobs. As each new requirement is released to the shop, your system's scheduling function should be able to tell you whether the start is at the proper point for normal production processing and completion. Be sure that order release is in control before measuring the plant on reacting to system priorities.

Eliminate any volume-based measurements (units produced, dollar-value produced, efficiency, utilization). These measurements do not relate to master schedule or customer service objectives. All measurements must be tied to priority or schedule. This is not an easy set of measurements to develop but it is the back-up detail that helps identify the contributors to your on-time completion record.

Classical input–output analysis is still a valid tool for monitoring work flow to be used in conjunction with your priority-based measurement. Input–output is a measure of the balance between the rate of introduction of work and the ability to complete it. It is best done at the facility (work center) level to help identify bottlenecks and constraining resources.

PAST DUE ORDERS

This can be applied to purchase orders, manufacturing orders (schedules), and customer orders and is only a very general measure of how things are going. As you gain control, the number of activities not completed on time should continue to go down. Other measurements are needed to help pinpoint the causes.

PARTS SHORTAGES BY REASON

MRP seeks to avoid shortages through planning—anticipate the need and take action before the shortage occurs. MRP assumes that the bills-of-material, inventory records, and on-order activities (due dates and quantities) are correct. Inaccuracy or failure to perform as expected will result in excess inventory or shortages. We compensate for known or assumed inaccuracies and uncontrollable factors with inventory buffers.

Shortages will happen. It is important to try to identify the cause of the shortage so that appropriate action can be taken to avoid its recurrence. Measuring shortage by reason gives us an indication of whether we are improving or not.

You must look at this measurement in conjunction with the next ones—measures of inventory level—to determine whether it is the buffers that are helping avoid the shortages or if you are, indeed, gaining control. Avoiding shortages is easy—just buy lots of inventory. Avoiding shortages through effective planning and control is the real goal.

INVENTORY MEASUREMENTS

To begin with, inventory must be measured by category—finished goods, raw materials, component parts—and in many cases by product group, by planner, or some other breakdown within each category. Second, inventory measurements, like most others, should be tracked as ratios to business volume or projected need. Third, it is trends that are important, not one particular measurement point.

For components and materials, measure the level relative to expected need based on the current plan. Expressed as "days supply," this is a measure of the amount of buffer inventory you are carrying. As you make efforts to reduce this buffer, carefully watch the number of shortages. As you gain control, you should be able to reduce both inventory and shortages at the same time.

NUMBER OF MRP EXCEPTIONS

MRP will recommend rescheduling of ongoing purchasing and manufacturing activities to better align them with the current needs. Assuming that the Master Schedule was put in place early enough (beyond CMLT) and that there are no changes to this Master Schedule or the bills-of-materials, the primary cause of exception messages will be "surprises"—unexpected shortages due to scrap and rejects, inventory or bill-of-material errors, and/or past-due activities. The number of exceptions, therefore, is a general measure of the amount of control.

You should watch exceptions by type (expedite, defer, cancel) and group them by the number of days of the recommended change. Watch for patterns which might lead you to the cause.

Exceptions are also caused by changes to the Master Schedule within CMLT. It is not always easy to isolate exceptions caused by schedule changes from those caused by scrap, rejects, and data inaccuracy.

ON-TIME SHIPMENTS (LATE SHIPMENTS BY REASON)

The ultimate measure of success is satisfying the customer; but, as indicated before, many companies ship on time primarily through expediting and extraordinary effort. Of course you will want on-time shipment, and a measure of success with MRP II is attaining and maintaining a high level of customer service, but you want to be sure that you did it through control, not heroic efforts.

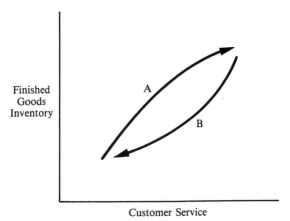

Fig. 11-1. Inventory investment versus service level.

Customer service can be increased easily by increasing finished goods inventory. I have seen a number of companies go back and forth along the paths indicated in Figure 11-1, increasing finished goods inventory which results in a better shipment record (line A). Eventually, someone will scream about too much inventory. Inventory is cut which results in a drop in service (line B). When the customer complaints get loud enough, inventory is increased and we're back to line A.

Back and forth these companies go, in a never-ending cycle. Perhaps what they don't realize is that there is a third factor in the equation that can be used to break this cycle. Figure 11-2 illustrates the three factors that are tied together such that you cannot change one without changing at least one of the others. You can go back and forth between inventory and service level or you can work with other combinations. If you can forecast better, for example, service can be improved without increasing inventory, or inventory can be reduced without affecting service.

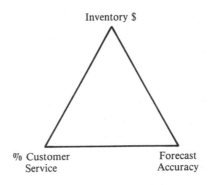

Fig. 11-2. The service–inventory–forecast triangle.

Another useful measurement, therefore, is forecast accuracy. A forecast can never be 100% correct. The less variation between forecast and actual demand, however, the better service you can achieve with less finished goods inventory. Another factor to watch is overall lead time. It is easier to forecast closer in than farther away. If you can reduce lead time, you will have more time to refine the forecast before firming up your plans within CMLT.

Measurements of System Performance

The system and its workings is not the objective and should not be allowed to overshadow the real reason you install systems: to manage information that supports the strategic goals of the company. Keep in mind, however, that the system only manages the information that it is provided and will accept and believe an untruth as long as it fits within the criteria the system expects. When you lie to the system, it will lie right back—at electronic speed. The information you get *from* the system is only as good as the data you put *into* it.

In line with the above statement of perspective, the following measurements are useful in establishing and maintaining the required levels of accuracy to support MRP II operations.

INVENTORY ACCURACY

Every MRP book, class, and magazine article ever written has specified a requirement for 95% inventory (record) accuracy or better for success with MRP. I don't know where that number came from originally, but it has become an accepted goal and is actually an aggressive but achievable objective.

It is important to define how this accuracy is to be measured. In most cases, when a total physical inventory is taken, the accuracy reported is total dollar value counted compared to dollar value according to the inventory records. This measurement benefits from the fact that errors are likely to be distributed both above and below the expected quantity, therefore, some of the error will be canceled by other errors.

For operational purposes, when you come up short on widgets, it doesn't help if there are extra gadgets you didn't know you had. For this measurement to be meaningful, then, it must be the percentage of items whose count is correct. In other words, if you were to count 100 items, we would expect 95 to be correct and 5 or fewer to have some error, without regard to how large or small that error happens to be.

Practically speaking, if an item is weigh counted or estimated because there are thousands of them on hand and they don't really matter that much, you can allow yourself a modest error tolerance of a few percent and still consider the count as correct. But for anything that is hand counted, there should be no acceptable level of error—the count is correct or not.

The only practical way to measure accuracy and achieve the goal of 95% or better is through an aggressive cycle counting program. Cycle counting allows you to count often enough to get meaningful data and trends. It also provides you with a realistic opportunity to discover the cause of errors (with a little detective work) and prevent their recurrence, which is the true key to improving and maintaining accuracy.

BILL-OF-MATERIAL ACCURACY

You knew this one would be next, didn't you? Right up there with inventory accuracy in every book, class, and magazine article is the "magic" figure of 98%+ for bills-of-materials. I won't bore you with a long, drawn-out justification for bill accuracy or how to achieve it with timely, accurate inventory transaction reporting. Suffice to say that this number is also aggressive but achievable and that your success with MRP depends to a great extent on how well you attend to these two basic pieces of information that are central to the calculations that MRP carries out.

DATABASE AUDITS

The typical MRP II database contains literally hundreds of pieces of information for each item that not only characterize the items and their processes but also tell the system how to manage that particular product, component, or material. Once you understand these codes and characteristics, there are undoubtedly many things that you can check for consistency and completeness using standard reports and/or database queries.

There is probably a code in the item file that specifies whether the item is purchased or manufactured. For a manufactured item, for example, you would expect to find materials specified in the bill, a routing to describe the process, planning controls that are appropriate for a manufactured item, and that the item might be stored in a particular area of the stock room, it might have classification codes (item type, item class, value class, etc.) that are specific for manufactured items, and on and on. It is a relatively simple job, in most systems, to do comparisons of related fields such as these using a "query" utility. Such audits should be part of regular, on-going efforts to maintain and improve data integrity.

It is important to look upon these database audits as an opportunity to make things better, not as a way to find out who's messing up. As with cycle counting, the objective is to identify errors *and the source of the error;* then correct the mistakes, of course, but more importantly address the source of error so that it will not recur. Try to avoid placing blame. Enlist the assistance of all those responsible for maintaining the database to self-police and self-correct errors. Look upon each identified error as a treasure you have just unearthed. The error was there before, you just didn't know it. Now that you know about it, and how to fix it, you are much better off and your system will be that much more useful.

Measurements of errors can work both ways. The greater number found, the more will be corrected and the better things will be. Finding fewer errors, however, could mean that there are fewer to find, or it could mean that you are not looking hard enough or people are not reporting the errors they have found. Be careful how you view the error rate and what incentives you apply to finding errors. Your response could easily motivate people to either hide errors or to create them just so that they can be found.

Continuous Improvement

Whatever measurements are put in place and whatever the current performance is for these measures, the primary goal should be to improve. Many companies find that the initial measurements can be somewhat embarrassing—nobody likes to admit to a 40% inventory accuracy. It is important not to be discouraged and not to blame the measurement. A poor measurement score is an opportunity for improvement, and often the poorer the score, the easier it is to generate dramatic improvement quickly, offering some strong positive feedback for your project. Also beware of "perfect" performance. Very few things are consistently perfect. Question the value of the measurement, the precision of the measurement tool, and whether you are measuring a meaningful result—whether workers are simply satisfying the measurement or actually achieving the desired outcome.

Remember that measurements are not very useful if they are kept secret. Open display of measurement results is one way to enlist the assistance and participation of company employees at all levels to drive improvements.

There are books and magazine articles that will specify desired levels for key measurements, such as 95% on-time shipment to customers or less than 5% variance of sales to forecast. I believe that the goals for each company should be set by the company and should be revised regularly as each goal is attained.

The initial goals should be aggressive but realistic. If your inventory accuracy is currently 40%, set a goal of 85% within three months, then 90% two months after that. When these goals are achieved, raise the target to 95%, then 97%. Some kind of recognition or rewards would be appropriate when intermediate goals are achieved.

Executive Information Systems (EIS)

One of the newer buzzwords made popular recently by software vendors is the Executive Information System (EIS). The idea behind EIS is that few executives have the time or the inclination to look at the truckloads of data in their corporate systems that describe the state of the business in detail. The executive, therefore, should be given access to summary information, in an easy-to-use format, that will help identify problem areas and monitor the overall health of various areas of the company. It is also desirable that the executive be able to dig down into successively more detailed levels of back-up information to investigate any item that captures his or her interest (referred to as "drilling down" in the EIS vocabulary).

An EIS is an extension of the existing information system. It adds no new information but merely offers convenient access to what is already there. An EIS is, by definition, an interactive facility with few preset formats. The executive is encouraged to explore the information and to trace relationships and causes as he or she sees fit. Today's offerings are usually graphically based using windows-like displays and mouse-driven controls.

Most EIS systems can import the retrieved data into spreadsheets, database products, and word-processing documents. Most also interface with electronic mail to facilitate communications with the functional managers.

When implementing an EIS, it is important to work with the IS staff as they know best what is on your system and how it must be accessed. You should also enlist the cooperation and assistance of the functional managers within the company. They will also benefit from the summarized data and, if they are not involved in the process, they may view the new application as a way for senior management to keep an eye on them, and they will be resistant.

The EIS is a response to "Information Overload," the overwhelmed feeling that many of us get when faced with a five-hundred page report jam packed with numbers. If we summarize and simplify, the areas of concern can be identified. Then, more detail can be retrieved in an orderly fashion and only as needed.

The People Side

A major impediment to implementation of any new management system is an employee's motivational system. People tend to do what is rewarding (pleasurable) and avoid what causes pain. The motivational system that is currently in operation in your plant is a reflection of the priorities and directions that currently exist in the company as well as a derivative of those from the past. When new systems and procedures are implemented, it is important to evaluate the motivational environment to ensure that the desired behavior is properly encouraged.

If your new information system is designed to provide plant-floor personnel with dynamically generated priorities tied to current demands and schedules, then it is important that production heed these priorities for the system to be effective. If these same production people are motivated based on the number of units produced per month or daily production counts, they are not tied in to the system's priorities. When it is time to select the next job from the available queue, he or she will select the one that provides the best result for his or her incentive system, not necessarily the one that the system says is most critical. He or she may well select the lowest priority job next because it gives him or her the best numbers for the day or month.

One of the least recognized and often most difficult areas of system implementation is identifying where employee motivation is not in line with the objectives of the new system, and correcting the motivational system. Often, this can spell the difference between mediocre results and truly outstanding improvements. Appropriate measurements, effectively tied to the employees' motivational system, are key to getting the results and benefits that your information management system can help you generate.

12. Planning for and Implementing MRP II

When Alice encountered the Cheshire Cat, she asked him which road she should take. The cat asked Alice, in turn, where it was she wanted to go. When Alice told the cat that she wasn't exactly sure, the cat replied that it really didn't matter, then, which road she took. Like Alice, many companies really don't know what their destination is. They have no overall plan for systems implementation, therefore, it's hard to tell if they are really making progress since there is no clear goal or path. Installing the system and the software is not a goal, it is a (minor) step along the way. Goals must relate directly to bottom-line business benefits to be valid, measurable, and justifiable.

The planning process must start at the top with overall company goals outlining what the business will be like in two years, five years, and beyond. In line with general business planning, there should be some clear statements about the projected size of the company in dollar sales, units, employees, and facilities. The plan should include projections of the kinds of products and their volumes that will produce the sales dollars and how, in general, they will be produced including the amount of subcontracting and the level of vertical integration anticipated. These estimates serve as a definition of the environment in which the overall company goals will be accomplished. Other specific high-level goals come next, as targets for detailed planning of the resources and facilities that will be required to support the projected levels of production.

Next comes a realistic assessment of the state of the business today. This includes the establishment of baseline measurements which will be very important later when you are trying to quantify the improvements and direct results of the project.

Armed with this background information, start drawing a line from point A to point B. Project the increase in business and identify the need for additional resources to meet those needs as they grow. Speculate how internal systems, as well as system-to-system links (within the company and with your trading partners) could help tighten management control, make operations more effective, eliminate waste (anything that adds cost but doesn't add value to the product), and help support more business with a given level of resources.

Looking back at the nonsystem issues addressed in this book, make a

PLANNING FOR AND IMPLEMENTING MRP II

list of areas in your company that can be improved through changes in procedures, different organization, education, discipline, and attention to detail. Put together a program to address these areas prior to, or along with, investing in any new technologies. Remember that any new equipment or software brought into the picture must be supported with adequate training for the users and an appropriate support structure (maintenance, continual training for skills enhancement and for replacements, new or revised procedures and controls, appropriate measurements).

Once you have developed a detailed plan, be sure that you put it in writing. The more formalized and detailed the written plan, the easier it will be to assess your progress, manage the project, and set priorities at each step of the way.

Even a written plan will not necessarily accomplish much unless each item on the plan has a name and a date assigned to it. Clearly defined responsibility, the proper authority to carry out the task, and a real sense of "ownership" by the responsible individuals are the prerequisites for successful accomplishment of the project objectives.

The detailed plan must also have intermediate goals and measurement points. You cannot lay out a two-year plan, for example, and come back two years later to see if you made it or not. Regular status checks (weekly, monthly) are needed to assess progress and provide further direction to keep things on track.

There will come a time in every project when there will be other priorities that jeopardize your schedule. Some tough decisions will be required—some uncomfortable tradeoffs. One thing to avoid, if at all possible, is allowing the schedule to "slip." This sends a clear signal to the participants that the project is not as important as they were led to believe, and further conflicts and slipping become inevitable.

When you set your original schedule make it aggressive but feasible. A schedule that is too lax will encourage procrastination and not achieve the improvements and payoffs as soon. A too-aggressive schedule will lead to frustration, protest, and probably failure. There's a fine line between too lax and too aggressive. The challenge is to find the right level and stick to it.

The participants, those who have their names next to the dates on the project plan, must agree with the objectives and schedules. Imposition of an aggressive schedule by executive fiat is not the way to instill a feeling of commitment and ownership. There should be room for some give and take in formulating the schedule, within the framework just discussed.

It is important to manage an implementation project throughout its life. Most successful implementation efforts have weekly status meetings for the team members to coordinate their activities, keep each other up to date, and review the plans for the coming weeks. Often, the project will stall due to unexpected difficulties or other priorities. With a weekly review, these problems can be identified and addressed before they cause serious schedule delays. Sometimes there will be temptation to cancel a weekly meeting because there is nothing to discuss. No progress during a week is definitely a situation to be addressed! There should be activity *every week*. The project cannot be allowed to rest, ever, if you wish to reach your goal.

Inevitably, there will be disappointments and "snags" in the project. Letting the schedule "slip" should be the last resort. As soon as a problem is identified, identify the solution and apply the resources necessary to get back on schedule. It is smart to plan a modest buffer into each major task's schedule, but keep the buffer at the end of the task, after the official due date, and plan not to use it if at all possible. If you don't need the buffer, you might just finish the project ahead of schedule. Wouldn't that be a nice gift to company management, not to mention a feather in the project team's collective cap?

One thing about systems projects like MRP II: there's always more that can be done. As your initial project nears completion, start planning the next project. After all, the key to survival in a highly competitive world is continuous improvement. Once you have assembled a team and built enthusiasm and a mechanism for change and improvement, harness those resources for additional benefits.

Finally, remember that the people issues are the most important part of any project involving change. No computer system, software package, organizational change, or capital investment will pay off unless the people affected by the change are ready for it, embrace and support it, and work with you to make it happen. Adequate education and/or training of all affected employees, not just the ones directly in contact with the new system, is vital to the success of your implementation.

Remember, also, that the need for education is not a one-time requirement. Your organization and its business will change. It is likely that the software you installed will evolve over time. You may install additional functions or add enhancements to your system. Continued education will help the users keep up with the latest tools you are making available to them.

In addition to all of the changes mentioned above, the people themselves will move around. Some will leave and their replacements must be trained. Others will be promoted or otherwise moved to new areas of responsibility. The organization itself will likely evolve as new management techniques are employed and individual responsibilities will shift accordingly. Don't assume that the initial round of training is sufficient until the next software change. Think in terms of continuous improvement of the skills of the people as well as the specific tools and techniques being applied.

The Keys to Success

There are three things that, more than any others, determine the level of success that can be achieved in any systems project, whether it involves the purchase and installation of new equipment, new software, or simply changes in procedures or organization. These three key elements have nothing to do with hardware or software selection, how advanced the technology, or how much money you spend. They are: senior management commitment, a team approach, and education.

MANAGEMENT COMMITMENT

Any improvement project will require some effort and some level of change, usually felt most acutely at the middle and lower levels of the organization. Often, extra effort is required up front: such as file building, parallel operation, learning new procedures, the inevitable "learning curve." These represent an investment to implement the new ways. The burden most often falls on the people who have the least time to spare and who typically had no involvement in the decision to make the change (this can be avoided!). These people will only participate wholeheartedly if it is made important to them. Clear, visible management support, commitment, and involvement in the project is the best way to ensure the proper emphasis, to keep the project going when things are busy or there are other demands on their time.

Commitment means more than a gratuitous statement at the first announcement meeting. Senior management (chief operating officer and functional area management: director of manufacturing, plant manager, director of engineering, etc.) must be *involved* and show an active interest.

I taught some initial classes for an MRP II project a few years ago, and at the first session, the program started with a video tape of the corporate CEO stating the direction and importance of the project. He clearly and eloquently stated that this was a major, strategic project for the company and that the company's very survival in the marketplace was dependent upon the success of this undertaking. I said to myself at the time "This company will do well, the three elements are right here. We have strong management direction, the team is assembled here for training, and they are all committed."

About two years later, I happened to run into one of the team members at a conference and I asked him how the project turned out. Despite the apparent good start, the whole effort stalled after a few months and failed to achieve its objectives. The team did their best, sufficient education was provided, but the management commitment ended with the video tape. When things got busy, there was no push to keep the project effort going. When tough decisions had to be made, the project lost priority because it didn't have the high-level backing it needed. A lot of money was spent. A lot of good people got frustrated and left the company.

I hate to include such a negative example but it is an important point which is best illustrated in the negative. Proper executive involvement, while an essential ingredient, doesn't necessarily announce itself as a determining factor when looking at the results. By all means, the executive can take his or her share of pride and credit for a job well done, but recognize that the "owners" of the project, that is, the users, are the real heroes. Be sure that there is plenty of praise and recognition distributed to the project team and the entire company.

THE TEAM

An MRP II project, by definition, cuts across functional areas. Often, it requires formerly uncooperative or sometimes downright hostile groups to share information, cooperate fully for the good of the company, and consider each others' needs on a level with their own. An integrated system project cannot be carried out by one department and imposed on another. All affected areas must be full participants and take a share of the ownership and responsibility.

The project team should include the people who are the least able to take the time. The key players in the department or area, those who can "take charge" and "make things happen," are the ones who can make the project a success because they are the ones who are probably keeping their departments going. Everyone knows who these people are and they earn the respect of their peers through their accomplishments. They are the leaders, even if not recognized as such by their job titles.

The project team is responsible for developing the schedule (with executive guidance and backing) and assigning responsible individuals and dates to each item. The team monitors progress, coordinates activity, and reports to senior management.

EDUCATION

The first job of the project team members is to educate themselves. They must understand the need for the project, the "why" (theory), the effort required, the expected results, and the impact the changes will have on the organization, the procedures and activities, and the people.

Education and training must then be extended to all levels of the company that are affected by the changes. In most cases, the improvement project will involve changes in procedures, disciplines, day-to-day activities and responsibilities for everyone in the company, whether or not they have direct contact with a system terminal. People have a natural tendency to resist change and fear the unknown. The best way to overcome these fears is through education.

How can you measure the effectiveness of your plan and your team? The proof is in the accomplishments and in the overall atmosphere in which they take place. Setting an aggressive plan and carrying it out is obviously the ultimate measure of success. The obvious things to watch are on-time completion of tasks, meeting intermediate objectives, and progress toward full implementation according to documented project goals.

A more subtle method, but just as important, is to watch the attitudes of the project participants and nonparticipants. Not everyone will wholeheartedly embrace the project and cooperate in its completion. You must identify those who are not a part of the team and determine why they are resisting. Many times people will either feel "left out" or they might be intimidated by the technology. Don't let these common challenges impede progress. Watch for them, detect resistance before it becomes a problem, and address it. Usually a little counseling and some training will bring these people into the fold.

Using Consultants

When dealing with new (to you, at least) technologies, it is often helpful to enlist the assistance of an expert in the field. A good consultant can help you understand and properly apply the technology and can help you avoid costly errors by virtue of his or her experience with other installations.

A consultant should not, however, be your project manager, run the implementation team, or take responsibility for the project. You cannot delegate an integration or systems implementation project outside of your company. Since the principal impact of these projects is on the culture and activities of the employees, it is the employees who must retain ownership of the changes.

The consultant should be used as a resource for technical information, for the benefit of his or her experience, and for his or her intelligence and insight. He or she can also provide an objective viewpoint, not colored by your company's past experiences or current culture and personalities.

Some consultants will contract for extended services over a long period of time, such as one or two days per week for a year. While some companies and some consultants are comfortable with this kind of relationship, I have always preferred a lower level of involvement. A typical consulting arrangement for me would include no more than several days per month and often only a few days per year.

I believe that there can definitely be too much of a good thing. If the consultant is to be on-site and available every week, even if only for a day or two, I believe the employees can become too dependent. Too often, the project doesn't really make much progress on the days that the consultant is not on-site, and the employees don't develop the independence and self-reliance that is necessary to achieve permanent change.

During critical implementation phases, you may need assistance for several days or weeks at a time, but it is important to preserve your self-reliance. Don't think of the consultant as a company resource—he or she is strictly an advisor and cannot be held directly responsible for any portion of the project or company operations.

Another outside resource that you may utilize is contract technical services. Programmers and engineers can be contracted on a project basis to accomplish defined tasks such as writing programs, installing networks and cabling systems, or interfacing equipment. These people or companies can and should be held responsible for their specific tasks within the project. The difference is that the tasks assigned are one-time efforts that have a defined result and end point. The expected results can be clearly specified in the contract, and accomplishment can be proven through testing of performance of the product.

When contracting for services, be sure that there is a measurable product. Agree on the expected result and how its completion is to be proven to the satisfaction of both parties. Check references and be sure that the contractor has demonstrated an ability in the area that you will be contracting.

In this chapter, we'll take a brief look into the future, tracing the trends of the recent past to their logical conclusions. As with any other forecast, there are bound to be deviations from the expected patterns as well as some surprises along the way. In general, however, I believe that the thoughts presented here are likely to come to fruition within the next few years.

There are three sections in this chapter. The first contains projections for manufacturing in general, and the other two look at trends in computer hardware and software.

13. Trends

Industry Trends

Manufacturing today, in nearly all markets, is far more competitive than ever before. Competition tends to be more global and promises to become even more so. With emerging countries rapidly developing manufacturing capabilities and the reduction of trade barriers worldwide, most manufacturers are now in global markets with global competition. Even a company that doesn't import or export is in a global market—like it or not.

At the same time, product cycles are getting shorter, at least in part due to the increased competition. Rapid advances in technology also render products obsolete far more quickly than before.

For these reasons and others, we no longer have the luxury of ample time to develop new products and bring them to market on a leisurely schedule. Anything that we can do to reduce design and engineering time, and get products to market more quickly, enhances flexibility and competitiveness.

Twenty years ago, the conventional wisdom was that higher volume results in lower unit cost. While the concept of economies of scale is still valid in itself, the realities of the market place definite limits on this logic. When considering fixed and variable costs, it is simple to see that when a larger production quantity absorbs a fixed cost, the impact on the individual units is reduced. If, however, a large number of the goods produced end up as obsolete inventory because of the introduction of an improved product or a lower-priced competitor, then the savings are illusory. The result is a trend toward smaller "economic" production quantities. This can be achieved only by reduction of fixed costs. Fixed costs include development, engineering, and production set-up costs. Anything that helps reduce these costs improves competitiveness by allowing us to meet market needs more precisely. Fortunately, the things that help reduce costs often reduce the time required as well, giving us more flexibility.

With smaller production lots, shorter product cycles, and intense competition, companies are striving to differentiate their own products from those of the competition. While the "economies of scale" theories of the past tended to reduce the selection of products available because the price of entry is high for high-volume production, the foregoing trends increase diversity. In an effort to avoid direct head-to-head competition, and taking advantage of the flexibility offered by the new technology, companies are now offering an ever-wider variety of products aimed at smaller "niche" markets. This often allows higher margins and discourages direct competition because of the limited sales potential of the precisely defined target customers.

All of these trends are related through the interaction of increased competition and changing market conditions combined with the advances in technology and management that allow the outlined changes in the production realities.

Manufacturing management has been constantly searching for new tools and techniques that will improve competitiveness, increase flexibility, reduce costs, and help manage resources more effectively. This search has led us through a bewildering array of new software products, theories, and systems, all designed to "save" existing manufacturers from the competition. Sometimes, it's difficult to distinguish between what is really new and different and what is just hype or merely a refinement of an existing technique.

This is especially true in the business and planning area, but applies in engineering and plant operations as well. For example, MRP II is a well-established approach to business planning and control. Several years ago, however, "Just-In-Time" was widely touted as the newest and best method for becoming and remaining competitive. People started to ask (and some are still asking): "Should we install MRP II or JIT?" or "Should we scrap our MRP system and use JIT instead?" In fact, JIT is not a system at all but a management philosophy that can be applied in any company, whether MRP is in use or not. Actually, MRP II is often used as one of the tools to achieve JIT objectives.

The problem is that people focused on the new acronym without understanding what the idea was all about. JIT is a philosophy of identifying wasteful practices and taking action to reduce or eliminate waste. MRP II, on the other hand, is a set of calculations, techniques, and information management tools that help us manage the situation as it is. There is no conflict and no either-or choice.

The real danger in the development and promotion of new techniques and theories is that the attention is drawn to the technique and distracted away from what is really important—producing a quality product on time and at an acceptable cost. In addressing any kind of system or theory, we must be aware of that danger and keep clearly in mind that no clever technique, fascinating theory, sparkling new computer, lightning fast communications system, or expensive new machine is worth anything unless it directly supports the objectives of the company and helps it to become and remain competitive and profitable.

A manufacturing company is an entity in which all of the parts are interdependent and which is in existence for a single reason: to make a profit by producing and selling a product.

Hardware and System Software Directions

The trend in computer systems (hardware) is up in capability and down in price, continuing in a direction that has prevailed for twenty years. The ubiquitous "personal" computer will grow more powerful each year, while last year's technology becomes a commodity product for sale at razor-thin margins by second-tier producers, often using significant content from third-world producers.

Although the trade press has declared that mainframes and mid-range systems have reached the end of their useful lives, as was the case with Mark Twain, the reports of their deaths are premature. Vendors and customers have found the larger systems to be very useful in a cooperative processing environment with PC's providing significant processing power at the desktop, reducing the demand on the centralized system which is focused on corporate data management, communications, and multiuser-oriented tasks.

The emergence of the new smaller and more powerful workstations, somewhere between the PC and the traditional mid-range, has somewhat confused market strategies and positioning in both markets. Large PC's compete directly with workstations in the scientific and CAD/CAM area while, at the same time, new business applications for the (mostly UNIX-based) workstations have the traditional mid-range vendors looking over their shoulders. In addition, several of the mid-range vendors have recently introduced mainframe versions of their systems. These latter introductions indicate two things. First, the mainframe is not dead. If savvy companies like DEC are willing to make the investment necessary to go after this business, there must be significant business potential there. Second, the trend of smaller systems growing into market segments previously dominated by larger systems is a characteristic of the industry as a whole, not just PC's moving into mid-range territory.

System connectivity is a major strategic direction for all systems in all markets. While the dominant vendors will continue to support and promote their proprietary protocols and standards, neutral, open alternatives will gain ground as well as built-in or bolt-on type adapters that allow easy connection to widely used communications links like Token Ring networks, Ethernet, and TCP/IP.

On the software side of interconnectivity, more vendors will offer translation capabilities to neutral standards like IGES or the upcoming STEP standard on the engineering side and emerging EDI standards for communication with trading partners. With more widespread development and acceptance of open protocols, information exchange both within the company and outside should become more routine.

Proprietary operating systems will always be with us but UNIX and its derivatives have shown surprising strength and longevity. Due, at least in part, to the increasing popularity of Reduced Instruction Set Computer (RISC) systems, more applications are becoming available for UNIX and its variants and this segment of the market will continue to grow. It is yet to be seen whether open systems will overtake proprietary ones in the business and planning (MRP II) area. So far, their penetration of this market is minimal.

Application Software

MRP II will remain the dominant business and planning system format for the foreseeable future. Vendors will continue to enhance and extend their packages in an effort to stay ahead of the competition and distinguish their product from the pack through special features. Interestingly, these efforts to enhance the products tend to make them more alike than different. Responding to competitive pressure causes vendors to duplicate any feature or function that the competition introduces, just to stay up-to-date.

An area of vigorous growth is the inclusion of "international" enhancement functions to existing packages. Most now offer alternate (language) descriptions, currency conversion, value added tax, and landed cost (import/export) features. Most systems will also continue to expand into peripheral areas such as distribution management and engineering release control.

New software vendors are expected to appear each year, as they always have, but most will never make a significant impact on the market in general. The successful ones will be niche players, focusing on a specific industry segment and building a reputation in a relatively small, defined market.

Alternatives to traditional MRP II will emerge but most likely will be unable to win over significant converts. Many of the existing "alternatives" are really traditional MRP II with a slight twist, and not really alternatives anyway. The one notable example of a true alternative that has been quite successful is the Marcam Prism "process model" which, since it is a patented technique, will not spawn the clones that MRP II has.

Cooperative processing approaches, including Executive Information Systems, will continue to find more widespread acceptance. The phenomenal success of Microsoft Windows in recent years has accelerated the development of graphical "front end" processors for use with existing MRP II systems and newly developed systems to provide easier access to the mountains of information contained in MRP II.

IBM's introduction in 1989 of its "CIM Architecture" and supporting products reflects a trend toward systemization of CIM that other vendors have also promoted. The tools and products encourage interconnection between individual resources, but the overall approach emphasizes some centralized controls and facilities as embodied in the CIM Architecture's repository (directory of data, its source, and usage) and data store (centralized storage function for shared information). Similar architectures and products can be expected from the other vendors of systems for manufacturers.

In services, systems integration contractors will continue to offer complete project management services (design, engineering, installation, etc.) for the larger companies. For smaller companies, the improved connectivity being built into new system products and the availability of "enablers," development tools that generate the interfacing programs with minimal technical knowledge required of the user, will bring do-it-yourself CIM connections within the reach of companies of all sizes.

Continued development of neutral standards and popularity of open systems will bring more compatible equipment and software to the market from an increasing variety of vendors. While standards and open systems are unlikely to ever completely displace proprietary products, it is likely that only the biggest vendors will continue to develop and support proprietary systems and protocols, and the smaller vendors will comply with either the open versions, the biggest proprietary ones, or both. Even IBM now offers products that work with open systems and neutral protocols, although it has not and most likely will not abandon its long-standing policy of setting its own standards.

14. Summary

There are literally hundreds of packaged MRP II software systems available on the market today, and just about every one does the "basic" functions in just about the same way. As you have seen, there are differences in the "bells and whistles" of various packages, and there are definitely differences in the needs of different industry segments. The truth of the matter is that every company has a somewhat different way of doing business, a unique twist on the standard process, or a distinctive mix of methods and products. Software vendors try to distinguish their packages from the competition by appealing to these differences.

Fortunately, many of these differences are variations on a theme, and standard MRP II functions, with some flexibility driven by market demands and built in by the vendors, satisfy most needs pretty well. Beyond the provided functions, there are ways to adapt (workarounds) to minor differences, enhancements available from third-party vendors that do not interfere with the basic package's functions, creative ways to use what's there, and smart (relatively low risk) ways to modify a package if none of the other solutions is acceptable.

Packages are here to stay and they offer a wealth of functionality for a small fraction of the price of custom code. An additional benefit is that a well-supported package will continue to grow and evolve as technology and management theories develop. It's a known fact that manufacturing is changing as the world changes, the economic situation fluctuates, third-world countries develop industrial capability, and developed countries continue to compete aggressively for the world's business.

Companies can no longer hide their heads in the sand, ignoring the world outside of their defined community. Every manufacturer is competing on a worldwide basis, whether they export and import or not. If you are not going after someone else's market "over there," be assured that "they" will be coming after your market here.

To stay competitive, indeed, to stay in business, you must adapt. You must be flexible to respond to changing conditions and changing market demands. Effective information management is a strategic weapon in this battle for survival and the sooner you realize this and take control of your information, the better prepared you will be to survive, and profit, in the near future and on into the next century.

Because packaged software vendors are also in a very competitive business, they too must change and adapt. You and your chosen vendor will be partners in the management of your company's information. Be sure to select a partner that is interested in the success of your business, and can provide the kind of support that you expect from a partner. Find one who is experienced in your area of business and can understand the challenges you face. Find a partner that is willing and able to work with you to implement your system and help train your employees. Find one that is likely to be there when you need support. Find one that is keeping their package up-to-date with a regular program of enhancements and upgrades.

Remember that there is no such thing as a perfect software product. All software has "bugs" and areas that can be improved. Assure yourself that your software partner will properly maintain their product and will be responsive to your calls for assistance.

Given that functional differences between packages are mostly in minor areas, such as a specific approach to a particular industry variation, and that you *will* have a choice of vendors and packages that will meet your requirements, the most important considerations for vendor selection must be in the support areas just listed. Make your "first cut" choice of software/vendor based on the availability of critical features. Make your final choice based on the support issues.

Success and Failure

I don't like to talk about failure of an MRP II installation, although I realize that there have certainly been some notable failures. The usual case, however, is not complete failure but rather not reaching the objectives originally envisioned. This gets a little tough to define because most implementation efforts don't clearly define the objectives, plan the project, and manage it to successful completion.

I believe that, most of the time, when a company agrees to spend a significant amount of money for computer hardware, software, and implementation effort, there is a general expectation of a return on that investment. Most often there is some sort of financial justification associated with the capital budgeting process, but that justification is too often vague and generalized. Any time the project team cannot document a real return of at least what was expected up front, the project could be considered a failure. Appropriate measurements, with adequate baseline readings before implementation begins, are critically important. You might want to review Chapter 11 for more thoughts on measurements.

The Oliver Wight Companies (Essex Junction, VT) did a survey a few years ago and received responses from over 1100 companies who had installed MRP II. While fewer than 10% rated themselves as "Class A"* users, two-thirds of the respondents said that the results equaled or exceeded their expectations. In fact, nearly all users (89%), no matter how "successful," said that they gained "better control of the business." The annual return on investment exceeded 50% for two-thirds of the companies surveyed and averaged 200% for the Class A users.

So, the success rate is really a matter of how you define success. Most companies making the MRP II investment get a good (financial) return on that investment, and nearly all gain better control of their business. Few reach the magical "Class A," but that is primarily a result of inadequate planning and management of the project. Generally, failure of the software to provide adequate functionality can be discounted as a reason for MRP II failure.

"The system," however, usually takes the blame. Many times, an unsuccessful project, or even a successful one that has failed to keep improving, is followed by the purchase and installation of a new system which may or may not be more successful than the first one. Here is my response to a letter written by a reader of the midrange manufacturing column that I write for *System 3X/400 Magazine* (reprinted with permission from Hunter Publishing, Des Plaines, IL).

Where Do We Go From Here?

A reader recently wrote to us with the following problem. His situation is that his company has used a packaged MRP II system for a number of years, and sixteen add-on applications have been developed (custom) which interface with the packaged system. Although they are several levels "down" with the packaged product—it is an older version of the package that has not been kept current with vendor-supplied updates because of the extensive modifications—they recently won a "Class A" user award.

Company management is trying to develop a strategic IS plan for the company's future. Since this MRP II system is a critical part of their operation, they are understandably concerned with its status, future, and what options are available. IS has determined, and has had confirmed by an outside source, that conversion of the custom subsystems to work with the latest version of the package would take 3,000 hours of programming time. They are also concerned that another 3,000 hours would be required for future upgrades as well.

There are concerns among the operational managers that a

---

*Class rating is developed according to answers to questions in an "ABCD Checklist" published by the Oliver Wight Companies.

conversion would cause disruption and may not be worth the effort (what would they gain?). Another manager is looking into PC-based MRP II systems. Yet another thinks that the company should fully commit to the current system, bringing in whatever resources are needed to maintain and build on the "custom" solution that is now in place.

Let me preface my comments with the admission that I am an unabashed believer in packaged software. Having been on all sides of the issue—user and seller of packages, developer and user of custom code, and support resource for packages, custom code, and modified packages—I have seen what can happen in all variations and I feel strongly that a packaged solution offers the best "bang for the buck."

The decision whether or not to upgrade or migrate from a heavily modified package, however, is as difficult as it is common. I offer the following comments and advice to this troubled reader based on ten-plus years of observing companies struggle with this issue.

The Devil You Know

With few exceptions, once a company has chosen a system and successfully implemented it, as this "Class A" user most surely has, the path of least resistance is to stay with the same package. Changing to any other package would require conversion of data and perhaps programs, retraining of the users, and general disruption of operations. Practically speaking, there is minimal difference between MRP II packages. Most have pretty much the same capabilities, despite the differences in screen layouts, database design, reports, etc.

If you are comfortable with your current vendor, believe that they will be able to provide adequate support, and feel that they are financially stable enough to survive in what has become a very competitive market, you have every reason to stay with that vendor. If the support is poor and the current system is lacking, look elsewhere. Otherwise, stay with "the devil you know."

Many companies change systems in an effort to "try again" after having failed in an implementation effort. In this case, a change of software can by psychologically (or politically) satisfying because it provides an opportunity to blame "the system" rather than the implementation effort. Often, the second effort succeeds, not because the software is better but because the implementers learned from their mistakes and did it right the second time around. This is not the case for our successful user.

I'm sure that I will get lots of letters from vendors for this next comment, but the truth is that an AS/400 model 50 application cannot be replaced by a PC network product—any PC network product. PC's were designed as single-user devices and will forever be limited by that basic design assumption. There are multi-user operating systems and utilities, network managers can do wonders with large numbers of users, and applications can be written for multiple access; but the fact remains that these uses are forced onto an architecture that was not designed for a multi-user environment and therefore cannot be as efficient or effective as a system designed from day one for this environment. This statement may not be true a few years from now (in the next generation of PC's) but it is true today. Forget the PC option.

Stay or Move Forward

The decision to stay with the old version or go through the effort and expense of upgrading to the new one is not as simple. In most cases of a migration to a newer version of the same product, there should be minimal operational disruption or need for extensive user retraining because the new version is a direct descendent of the old one. The biggest problem this user faces is the need to convert the custom subsystems and modifications.

The first rule of happy packaged software use is to avoid modification if at all possible. My experience has been that the vast majority of modifications to packaged software could have been avoided by better education in the functions and use of the package, creative use of the functions provided (workarounds), and changes to operational procedures to do things in a more "standard" way.

Admittedly, some earlier versions of current packaged systems lacked functionality in key areas and many users developed their own add-ons to compensate. The good news is that many of these modifications will probably not be needed with the new version because they will have been obviated by enhancements in the package.

A user considering an upgrade must put the effort into a detailed study to identify the purpose of each custom subsystem or modification, whether that purpose is valid (has the need gone away or was it really needed in the first place?) and whether the new version of the package satisfies the need, or whether a procedural change in conjunction with the new package will suffice.

This is work, especially if your documentation is less than complete, but it certainly makes more sense than converting pro-

SUMMARY

grams unnecessarily and perpetuating the problem in the case of future upgrades.

Generally, upgrading to the latest release is a good decision. Vendors cannot support "down level" code (especially if modified) as well as the current version, and most do not even attempt it. The newer code will undoubtedly have many enhanced functions that will allow your company to grow and improve its information control and integration. All of the major vendors are making significant enhancements to their products today to utilize cooperative processing and Graphical User Interfaces (see the May column in this space) and most are working very hard on CIM interfaces to CAD/CAM and shop-floor systems.

And let's face it, your business is changing whether you want it to or not. The marketplace is constantly changing, technology is rushing forward, and your company must continually adapt to the new realities if it hopes to survive and thrive. Packaged MRP II is also growing and changing with the needs of manufacturing, and the only way you can take advantage of these developments is to move to the newer versions of the package as they become available.

Custom modifications will continually hamper your ability to move in these new directions. There's no better time than now to start an aggressive program of eliminating custom modifications to your package. Use the prospect of an upgrade to the new version of your packaged system as the motivator for getting this effort off the ground. You will be glad you did, not only next fall or next year when you do complete your upgrade, but for years to come as you follow your software vendor into the future of manufacturing information management.

(end of article reprint)

While this company is certainly not a "failure" with MRP II, it is still facing the decision of what to do next, and it is likely that any perceived shortcomings within the company will be blamed on system deficiencies.

Most people have seen the humorous sign showing the "Phases of a Project" listed below:

1. Exultation
2. Disenchantment
3. Search for the Guilty

4. Punishment of the Innocent

5. Distinction for the Uninvolved.

Laugh, but not too hard. There is more truth here than we might like to admit. When the going gets tough, people tend to run for cover and look for someone or something to blame. "The System" is an easy target.

Remember that the functionality of the software itself is almost never the cause for failure to achieve your MRP II goals. I have seen companies with the best, most appropriate software fall flat on their faces, while other companies with poorly designed or inappropriate software succeed despite the limitations of their systems.

Appropriate software function can certainly make it easier to succeed. Today's MRP II products offer many features that address different industry needs, and that were not available even a few years ago. The competitive situation being what it is, we can look forward to continuing growth in function and enhanced features from all of the major package vendors in the future.

Glossary

ABC Analysis Application of the 80/20 rule that 80% of your costs or problems or sales are associated with 20% of your items or products or customers. Items, customers, etc., are assigned classification codes accordingly.

ABC (Activity Based Costing)—The identification and assignment of indirect costs to the activities that most directly benefit from them.

Application A packaged set of software programs that addresses a single functional area. Also called a "module." Applications are packaged according to the dictates of marketing considerations as much as for functional purposes.

Artificial Intelligence (AI) Software that attempts to emulate human decision making. Often rule-based "expert systems."

Buzzword An acronym or technical term, the use of which implies knowledge or understanding of the technology involved.

CAD/CAM/CAE Engineering applications: Computer Aided Design (or drafting or drawing), Computer Assisted Manufacturing, Computer Aided Engineering.

Campaign Scheduling Flow production in which a quantity of one item is produced, then a quantity of another, etc. The opposite of mixed-model scheduling.

Capacity Requirements Planning An MRP II application that gathers capacity and load information for plant facilities from other applications (shop activity control, MRP) and presents overload and underload information to the user for analysis and resolution.

Cell A group of facilities (dissimilar machines) arranged together for continuous production.

Cellular Production Production taking place in cells.

CFM Continuous Flow Manufacturing (see Flow Production).

CIM Computer Integrated Manufacturing.

CMfgLT (Cumulative Manufacturing Lead Time) The total time required to produce a product *not* including material acquisition. Production lead time at all levels within a product structure.

CMLT (Cumulative Material Lead Time) The total time required to produce a product including acquisition of all materials and completion of all production processes.

Continuous Production See Rate-Based Production, Flow Production.

CQLT (Customer Quoted Lead Time) The interval promised for shipment of a product to a customer.

CRP Capacity Requirements Planning.

Cycle Time (in a continuous process) The interval of time between the beginning (or end) of processing one item and the beginning (or end) of processing the next.

Discontinuous Production Order-based production in which work moves through the process in a group (job or work order). In most cases, all members of the group will be completed at one facility (job step or operation) and the entire group will move as one to the next operation.

DFM (Design for Manufacturability) The concept of including production considerations in the design process to produce a design that is compatible with the manufacturing capabilities of the plant. Also saves time in the design-to-production process.

Early Manufacturing Involvement (EMI) See DFM.

Executive Information System (EIS) A software utility that allows an executive to view critical strategic information from the corporate computer system in the form of easy-to-access tables and charts.

Finite Loading A scheduling philosophy in which capacity limits are recognized and respected.

Finite Scheduling See Finite Loading.

Flow Production Manufacturing environment in which material (work) moves from one activity or facility to the next individually. (See Rate-Based Production.)

Flow Time (in a continuous process) The length of time it takes for a single part or product to move from the beginning of a process (production line) to the end.

Group Technology (GT) An organized system for characterizing manufactured parts according to physical characteristics (such as size and general shape) and the processes required to manufacture them. Used to facilitate the retrieval of information (drawings, definitions, bills-of-materials) for same-as-except definition purposes and also for "design for manufacturability" tasks especially in conjunction with cellular manufacturing and process planning software.

Hardware Computer facilities including all of the physical equipment and excluding software.

Infinite Loading A scheduling philosophy which does not recognize capacity limitations.

JIT (Just-In-Time) See Chapters 1 and 4.

Job Shop Manufacturing company that is generally a subcontractor making products or components for another manufacturer.

KANBAN A Japanese word meaning "card." Name for an inventory or production management technique using a physical signal such as a card or empty container to trigger replenishment activity.

Master Production Schedule High-level production plan usually including all sellable items. The "driver" for MRP. Also called Master Schedule.

Mixed-Model Scheduling Flow production in which different products are intermixed. The opposite of campaign scheduling.

MMAS (Manufacturing Management and Accounting System) A term used in the government contracting arena.

Modular Bill-Of-Materials A form of bill-of-material that includes selectable options structured under a "feature" item, used for planning assemble-to-order products.

Module See Application.

MPS Master Production Schedule.

MRP Material Requirements Planning.

MRP II Manufacturing Resource Planning.

Operating System Computer software that provides basic system management functions.

Order-Based Production See Discontinuous Production.

Planning Bill-Of-Material Products or major subunits structured as a bill-of-material under a fictitious parent item using a forecast of expected relative usage as the quantity-per. Used for planning the components and materials of a family of products particularly when the exact products are not forecastable, such as in make-to-order situations.

Process Manufacturing Manufacturing which involves liquids and powders, usually by mixing, blending, heating (cooking), or through chemical reaction.

Production Line A continuous flow production facility consisting of a number of workstations.

Rate-Based Production Same as continuous or flow production, in which work moves from facility to facility individually rather than as a group. Scheduling of rate-based production is by item, by line, by day, rather than in identifiable work orders or jobs.

Release The initiation of a production or purchasing activity.

Resource Requirements Planning (RRP) A capacity checking function associated with production planning (for product families).

Rough-Cut Capacity Planning (RCCP) A capacity checking facility associated with master scheduling.

SKU (Stock Keeping Unit) An identified item in inventory. Equivalent to a part number or item number. Usually applies to finished goods inventory and is a common term in retailing.

Software Computer programs.

Subsystem An identifiable part of a system. Sometimes used to refer to an application module or a functional part of a module, e.g., the inventory subsystem of MRP II or the cycle counting subsystem of the inventory management application.

System A narrow definition is a computer and associated software and peripherals. A broad definition can include any set of procedures and environment for accomplishing a task. There can be manual "systems" for inventory control, for example, that consist solely of people and a card file.

WCM World Class Manufacturing.

Work Center A production facility most often associated with discontinuous production. Work centers are usually individual machines, groups of similar machines, or a team of people performing a specified activity.

Workstation A production facility within a production line. The equivalent of a work center most often used in continuous processing. Also a class of computer hardware often used for CAD/CAM and other processing-intensive applications.

Bibliography

Blackburn, Joseph, Time Based Competition—The Next Battleground in American Manufacturing, Business 1 Irwin, 1990

Goddard, Walter et al., ABCD Checklist, Oliver Wight Limited Publications, 1988

Goddard, Walter, Just-In-Time—Surviving by Breaking Tradition, Oliver Wight Limited Publications, 1986

Goldratt, Eliyahn and Jeff Cox, The Goal, North River Press, 1984

Gray, Christopher, The Right Choice—A Complete Guide to Evaluating, Selecting & Installing MRP II Software, Oliver Wright Limited Publications, 1987

Hall, Robert, Attaining Manufacturing Excellence, Dow Jones Irwin, 1987

Landvater, Darryl and Christopher Gray, MRP II Standard System, Oliver Wight Limited Publications, 1989

Maskell, Brian, Performance Measurement for World Class Manufacturing, Productivity Press, 1991

Mather, Hall, Competitive Manufacturing, Prentice-Hall, 1988

Plossl, George, Production and Inventory Control: Principles and Techniques, 2nd ed., Prentice Hall, 1985

Stalk, George and Thomas Hout, Competing Against Time, Free Press, 1990

Turbide, David, Computers in Manufacturing, Industrial Press, 1991

Wight, Oliver, The Executive's Guide to Successful MRP II, Oliver Wight Limited Publications, 1985

Wight, Oliver, Manufacturing Resource Planning: MRP II, Oliver Wight Limited Publications, 1984

Index

Accuracy, 23, 87
Active Ingredient, 102
Activity-Based-Costing (ABC), 29, 110–117
Actual Costing, 59
Aerospace, 10
APICS, 1, 4, 5
Artificial Intelligence, 9, 74
Assemble-to-Order, 71–74
Availability, 18, 22, 26
Available-to-Promise, 27, 28, 76

Back flushing, 55, 56, 59, 60, 103
Bar Code Data Collection, 120
Batch bill-of-material, 99
Batch formulas, 99
Batch-Lot Traceability, 104
Batch-Mix manufacturing, 51
Benchmarking, 33
Bills of Material, 5, 18, 19, 20, 26, 36, 37, 44, 46, 70, 79, 81, 85, 87, 96, 100, 115, 150
Blended Demand, 77
Borrow-Payback, 93
BPCS, 54

Build-to-Order, 6
By-products, 100

CAD, 121
Campaign Scheduling, 57
Cancel (recommendation), 23, 27
Capacity Requirements Planning (CRP), 24, 27, 37, 38, 40, 42, 44, 84, 89
Cell, 4, 46, 49
Changeover, 53, 54, 57
Chemical, 104
CIM, 31, 33, 97, 118–130, 165
CIM Architecture, 97, 129, 165
Closed-Loop system, 29
Commingled Inventories, 93
Compensating Ingredient, 102
Configuration Management, 9, 71, 72, 96
Configurators, 73
Consultants, 9, 159
Continuous Improvement, 30, 34, 46, 88, 151, 156
Continuous Flow Manufacturing (CFM), 51
Continuous Production, 5, 35

179

INDEX

Contract MRP, 84, 94
Cooperative Processing, 127, 163, 165
Co-products, 100, 103
Costing, 28, 92, 105
Critical Path, 64
Critical Resources, 58
C/SCSC, 97
Cumulative Material Lead Time (CMLT), 63, 74, 76, 147, 149
Cumulative Manufacturing Lead Time (CMfgLT), 64
Custom Manufacturing, 5, 26, 41
Customer Quoted Lead Time (CQLT), 70, 74
Customer Service, 65, 145, 148
Cycle Counting, 91, 150
Cycle time, 51, 53, 58, 60

DBMS, 123, 129
Defense (Department), 10
Defer (recommendation), 23, 27
Departmental cost(ing), 52, 59, 105
Department of Defense, 86
Descriptive Naming, 73
Design for Manufacturability (DFM), 51, 124
Digital Equipment Corp. (DEC), 163
Dimensioned Products, 73
Discontinuous production, 4, 35, 47
Discrete Manufacturing, 6, 7
D:P Ratio, 63
Drivers, 112
DX Priority, 96

Early Manufacturing Involvement (EMI), 51, 124
Economic Order Quantity (EOQ), 15, 23, 87
EDI, 164
Education, 8, 9, 12, 33, 137, 155, 156, 157, 158
Efficiency, 145, 146
Electronics, 10
Enabler, 129, 165

Ethernet, 164
Exception Messages, 147
Executive Commitment, 8
Executive Information System (EIS), 152, 165
Expedite (recommendation) 23, 27
Expediting, 38, 75, 76, 145
Expert Systems, 9

Features and Options, 71, 72
Federal Acquisition Regulations (FAR), 86
Feeder lines, 31, 47, 56
File Maintenance, 26
Finite Loading (scheduling), 9, 11, 38, 40, 41, 45, 59, 84
Flow Manufacturing, 5, 44, 49–62, 103
Flow time, 53
Focused Factory, 4
Food, 104
Forecast, Forecasting, 6, 17, 18, 27, 64, 65, 70, 72, 78, 89, 103, 148
Formulas, Formulations, 98
Full Pegging, 84

Government Contracting, 85–97
Grading, 98, 103
Granularity, 44, 54
Greater Demand, 77
Gross Requirements, 18, 22
Group Technology (GT), 81, 125

IBM, 54, 97, 127, 128, 165
IGES, 164
Independent Demand, 17
Infinite Loading (scheduling), 27, 36, 37
Inventory, 13, 26, 28, 46, 61, 63, 75, 89, 147, 148
Inventory Accuracy, 58, 149

Japan, 30
Job Shop, 2, 3, 4, 40, 79
Just-In-Time (JIT), 20, 29, 30, 33, 45, 49, 51, 55, 56

KANBAN, 29, 30, 31, 47, 51, 55, 56

Labor Costs, 60, 110
Lead time, 18, 19, 20, 23, 24, 32, 40, 47, 49, 51, 53, 64, 70, 85, 146, 149
Line Costing, 60
Line setting, 51, 55
Lot-for-lot, 23
Lot sizing, 18, 23
Low-Level Code, 23

Mainframe Computers, 119, 163
Make-to-order, 3, 5, 27, 69–71, 79–85
MAPICS/DB, 54, 56
Marcam Corporation, 106, 165
Master Schedule, 18, 28, 38, 40, 41, 56, 77, 87, 144, 145, 147
Master Scheduling, 27, 35, 42, 58, 63, 66–68
Material Handling, 61
Mating Parts, 85
Measurements, 140, 143–153, 154
Micro-computers – see PCs
Microsoft Windows, 127, 165
Mini-computers (mid-range), 10
Mixed Model Scheduling, 57
Modifying Software, 131, 139–142, 168–172
Modules, modular software, 25
Modular Bill-of-material, 71, 72
MRP. MRP II, 1–178
MRP III, 33
Netting, 22
Numerical Control, 120

Oliver Wight Companies, 167
Operating Systems, 32
Operation Costing, 60
Order-Based Production, 4, 6
Order point, 13–17, 30
OS/2, 127
Overhead, 60, 110
Overlap, 53, 54
Ownership, 9, 23, 29, 34, 155, 159

Paint, 104
Pegging, 84
PCs (Personal Computers), 10, 27, 45, 47, 54, 97, 119, 127, 163
Phantom Items, 99
Pharmaceutical 10, 104
Physical Inventory, 149
Pick list, 56
Planning Bill-of-Material, 70, 72, 80, 103
Plant layout, 4, 5, 49
Pools, 112
Potency, 98, 102
Priorities, 24, 36, 38, 43, 47, 146, 155
Prism, 106, 165
Process Control, 61, 105
Process Costing, 60, 105
Process Manufacturing, 6, 7, 10, 12, 51, 98–109
Process Model, 106, 165
Production Activity Control, 26, 35–48, 83, 106
Production Planning, 28, 58, 72, 89, 105
Productivity, 111
Pull signals, 51, 55
Pull systems, 19, 31
Purchasing, 26, 31
Push systems, 19

Quality, 46, 61, 105, 113, 130, 145
Queue, 39, 40, 47, 51, 53

Rate-Based Production, 6, 35
Resource Requirements Planning (RRP), 27, 28, 58, 89, 105
Request for Proposals (RFP), 132
Revision Level, 96
Rework, 60
RISC, 164
Rough-Cut Capacity Planning (RCCP), 27, 28, 38, 40, 58, 89, 105
Routings, 5, 26, 36, 72, 79, 81, 101, 115

Safety stock, 16, 65, 69, 77
Scheduling, 36, 54

Scrap, 59, 60, 147
Seasonal Demand, 67
Selecting Software, 131
Service level, 17, 66, 77
Shelf Life, 105
Ship-from-Stock, 6, 27, 68
Shortages, 146
Sequencing, 44
Simulation, 43
Simultaneous Engineering, 51
Standard Batch Quantity (SBQ), 99
Statistical Process Control, 46
Statistical Quality Control, 47
STEP, 164
Synchronous Production, 55, 56
Systems Application Architecture (SAA), 128
System Software Associates (SSA), 54
Systems 3X/400 Magazine, 168

TCP/IP, 164
Teams, 8, 158

Ten Key Elements, 86
Token Ring, 164
Toyota, 30, 55
Trends 161–165
Two-bin order point, 30

UNIX, 163
Utilization, 145

Variances, 59, 104, 115
Vendor Performance, 145

Waste, 30, 46
Windows, 127, 152, 165
WIP Inventory, 60, 75
World Class Manufacturing, 32
Work Breakdown Structure (WBS), 95
Work centers, 26
Work stations, 26
Workstation computers, 119, 163

Yield, 98, 101